U. Gresser (Ed.)

# Molecular Genetics, Biochemistry and Clinical Aspects of Inherited Disorders of Purine and Pyrimidine Metabolism

With Contributions by R. A. De Abreu, J. Aimi, F. X. Arredondo-Vega,
B. A. Barshop, M. T. Bausch-Jurken, F. Van den Bergh, G. Van den Berghe,
P. Casaer, S. Chaffee, J. Chen, P. De Cock, B. L. Davidson, P. M. Davies,
J. E. Dixon, J. A. Duley, W. Friedrich, B. S. Gathof, U. Gresser, M. Gross,
W. Gutensohn, W. Hartmann, J. F. Henderson, M. J. Henderson,
M. S. Hershfield, J. Jaeken, I. Kamilli, D. K. Mahnke-Zizelman,
F. A. Mateos, B. S. Mitchell, J. G. Puig, B. J. Roessler, R. L. Sabina,
A. S. Sahota, I. Santisteban, Y. S. Shin, H. A. Simmonds, P. J. Stambrook,
R. L. Stone, J. A. Tischfield, M. Tuchman, M. F. Vincent, D. R. Wagner,
K. Ward, R. W. E. Watts, H. Zalkin, N. Zöllner

With 40 Figures and 41 Tables

Springer-Verlag Berlin Heidelberg New York London
Paris Tokyo Hong Kong Barcelona Budapest

Privat-Dozentin Dr. Ursula Gresser
Medizinische Poliklinik der Universität München
Pettenkoferstr. 8a
80336 München
FRG

ISBN-13:978-3-642-84964-0     e-ISBN-13:978-3-642-84962-6
DOI: 10.1007/978-3-642-84962-6



Typesetting: FotoSatz Pfeifer GmbH, Gräfelfing

2127/3130-543210 – Printed on acid-free paper

*To Prof. Dr. Nepomuk Zöllner*
*with respect and gratitude*

# Preface

Inborn errors of purine and pyrimidine metabolism belong to the most interesting fields in clinical and experimental research. It was an inborn error of purine metabolism, which for the first time was treated with molecular genetic therapy in humans. Gene replacement therapy for ADA-deficiency became feasible in 1984–1985, soon after the cloning of the ADA cDNA and following the development of suitable retroviral vectors. In 1986 two girls with ADA-deficiency were the first humans, who were treated with gene therapy. This PEG-ADA was successful, up to now about 30 patients with ADA-deficiency have been treated in this way. In 1990 PEG-ADA was approved for treatment of ADA-deficiency by the US Food and Drug Administration.

This book summarizes the results of the "2nd Münchener Adventssymposium on Purine and Pyrimidine Metabolism", held in December 1991 at the Medizinische Poliklinik of the University of Munich. The aim of this symposium was to discuss the state of research in molecular genetics, biochemistry and clinical aspects of inherited disorders of purine and pyrimidine metabolism with special reference to diagnosis and treatment. We hope to convene scientists from all over the world to our annual "Münchener Adventssymposium on Purine and Pyrimidine Metabolism" in München also in the future.

Munich, June 1993                                          *Ursula Gresser*

# List of Contributors

Dr. R. A. De Abreu
Academisch Ziekenhuis Nijmegen, Sint-Radboudziekenhuis, Geert Grooteplein
Zuid 20, Postbus 9101, 6500 HB Nijmegen, The Netherlands

Dr. J. Aimi
Department of Biochemistry, Purdue University, West Lafayette, Indiana 47907,
USA

Dr. F. X. Arredondo-Vega
Departments of Medicine and Pediatrics, Duke University Medical Center,
Durham, NC 27710, USA

Dr. B. A. Barshop
Department of Pediatrics, University of California, San Diego, La Jolla, CA
90293, USA

Dr. M. T. Bausch-Jurken
Department of Cellular Biology and Anatomy, Medical College of Wisconsin,
8701 Watertown Plank Road, Milwaukee, Wisconsin 53226, USA

Dr. F. Van den Bergh
Laboratory of Physiological Chemistry, International Institute of Cellular and
Molecular Pathology, UCL 7539, 1200 Brussels, Belgium

Professor Dr. G. Van den Berghe
Université Catholique de Louvain, Faculté de Médecine, Laboratoire de Chimie
physiologique, Avenue Hippocrate 75, UCL 7539, 1200 Bruxelles, Belgium

Dr. P. Casaer
Department of Pediatrics, University of Leuven, 3000 Leuven, Belgium

Dr. S. Chaffee
Departments of Medicine and Pediatrics, Duke University Medical Center,
Durham, NC 27710, USA

Dr. J. Chen
Department of Medical and Molecular Genetics, Indiana University School of Medicine, Indianapolis, IN 46202, USA

Dr. P. De Cock
Department of Pediatrics, University of Leuven, 3000 Leuven, Belgium

Dr. B. L. Davidson
5520 MSRBI Box 0680, University of Michigan, 1150 West Medical Center Drive, Ann Arbor, MI 48109-0680, USA

Dr. P. M. Davies
Purine Research Laboratory, Guy's Hospital, London SE1 9RT, UK

Dr. J. E. Dixon
Department of Biological Chemistry, University of Michigan Medical School, Ann Arbor, Michigan 48109-0606, USA

Dr. J. A. Duley
Purine Research Laboratory, Guy's Hospital, London SE1 9RT, UK

Priv.-Doz. Dr. W. Friedrich
Universitäts-Kinderklinik, Prittwitzstraße 43, 89075 Ulm, FRG

Dr. B. S. Gathof
Medizinische Poliklinik der Universität München, Pettenkoferstraße 8 a, 80336 München, FRG

Priv.-Doz. Dr. U. Gresser
Medizinische Poliklinik der Universität München, Pettenkoferstraße 8 a, 80336 München, FRG

Priv.-Doz. Dr. Dr. M. Gross
Medizinische Poliklinik der Universität München, Pettenkoferstraße 8 a, 80336 München, FRG

Professor Dr. W. Gutensohn
Institut für Anthropologie und Humangenetik der Universität, Abteilung Biochemische Humangenetik, Goethestraße 31, 80336 München, FRG

Dr. W. Hartmann
Department of Pediatrics, University of Ulm, 89075 Ulm, FRG

Professor Dr. J. F. Henderson
Department of Biochemistry, 4-74 Med Sci Bldg, University of Alberta, Edmonton, Alberta T6G 2H7, Canada

Dr. M. J. Henderson
Department of Chemical Pathology, St. James's University Hospital, Leeds, UK

Professor Dr. M. S. Hershfield, M.D.
Box 3049/Room 418 Sands Building, Duke University Medical Center, Durham, North Carolina 27710, USA

Professor Dr. J. Jaeken
Department of Paediatrics, University of Leuven, University Hospital Gasthuisberg, Herestraat 49, 3000 Leuven, Belgium

Dr. I. Kamilli
Medizinische Poliklinik der Universität München, Pettenkoferstraße 8a, 80336 München, FRG

Dr. D. K. Mahnke-Zizelman
Department of Cellular Biology and Anatomy, Medical College of Wisconsin, 8701 Watertown Plank Road, Milwaukee, Wisconsin 53226, USA

Dr. F. A. Mateos
Devision of Clinical Biochemistry, Hospital General, Pl 10ª, Hospital "La Paz", Paseo de la Castellana, 261, 28034 Madrid, Spain

Professor Dr. B. S. Mitchell
Department of Pharmacology and Internal Medicine, CB 7365 FLOB, University of North Carolina at Chapel Hill, Chapel Hill, North Carolina 27599-7365, USA

Professor Dr. J. G. Puig
Servicio de Medicina Interna, Hospital General, Pl 10ª, Hospital "La Paz", Paseo de la Castellana, 261, 28034 Madrid, Spain

Dr. B. J. Roessler
5520 MSRBI Box 0680, University of Michigan, 1150 West Medical Center Drive, Ann Arbor, MI 48109-0680, USA

Professor Dr. R. L. Sabina
Department of Cellular Biology and Anatomy, Medical College of Wisconsin, Milwaukee, Wisconsin 53226, USA

Dr. A. S. Sahota
Department of Medical and Molecular Genetics, Indiana University School of Medicine, Indianapolis, IN 46202, USA

Dr. I. Santisteban
Departments of Medicine and Pediatrics, Duke University Medical Center, Durham, NC 27710, USA

Priv.-Doz. Dr. Y. S. Shin
Dr. von Haunersches Kinderspital der Universität München, Lindwurmstraße 4,
80336 München, FRG

Dr. H. A. Simmonds
Purine Research Laboratories, Guy's Tower (17th & 18th Floors), Guy's Hospital,
London Bridge, SE1 9RT, UK

Professor Dr. P. J. Stambrook
Department of Anatomy and Cell Biology, University of Cincinnati College of
Medicine, Cincinnati, OH 45267, USA

Dr. R. L. Stone
Department of Biological Chemistry, University of Michigan Medical School,
Ann Arbor, Michigan 48109-0606, USA

Professor Dr. J. A. Tischfield
Department of Medical and Molecular Genetics, Indiana University School of
Medicine, Indianapolis, IN 46202, USA

Dr. M. Tuchman
University of Minnesota School of Medicine, Department of Pediatrics and
Clinical Pharmacology, Minneapolis, Minnesota 55455, USA

Dr. M. F. Vincent
Laboratory of Physiological Chemistry, International Institute of Cellular and
Molecular Pathology, UCL 7539, 1200 Brussels, Belgium

Dr. D. R. Wagner
Medizinische Poliklinik der Universität München, Pettenkoferstraße 8 a, 80336
München, FRG

Dr. K. Ward
Airedale General Hospital, West Yorkshire, UK

Professor Dr. R. W. E. Watts
Wellington Hospital, Wellington Place, London NW8 9LR, UK

Dr. H. Zalkin
Department of Biochemistry, Purdue University, West Lafayette, Indiana 47907,
USA

Professor Dr. N. Zöllner
Medizinische Poliklinik der Universität München, Pettenkoferstraße 8 a, 80336
München, FRG

# Contents

# I Purine Salvage Enzymes

# 1 A   Hypoxanthine Guanine Phosphoribosyl-transferase (HGPRT) Deficiency

# 1   Introductory Remarks

R. W. E. WATTS

It is my great pleasure and privilege to chair the first session of the 2nd Munich Advent Symposium on Purine and Pyrimidine Metabolism. This is a field in which the work of the Medical Polyclinic of the University of Munich is well known and highly regarded worldwide. The two inborn errors of metabolism involving the purine salvage metabolic pathways are hypoxanthine guanine phosphoribosyl-transferase (HGPRT) deficiency and adenine phosphoribosyltransferase (APRT) deficiency. The range of clinical phenotypes associated with HGPRT deficiency is considerably wider than that observed in APRT deficiency. A considerable number of different qenomic lesions have now been identified in patients with HGPRT deficiency. These comprise the usual range of point mutations, single base deletions with frameshifts, small and large deletions, and insertions. It has not proved possible, so far, to clearly correlate these changes with either the clinical phenotypes or with the substrate binding sites on the enzyme molecule.

In my opinion, the term "Lesch-Nyhan-syndrome" should be strictly reserved as an eponym for those cases with fully meet criteria defined by Lesch and Nyhan. However, it should be emphasized that a patient whom we would now diagnose as having the Lesch-Nyhan syndrome was reported at an earlier date by Cattell and Schmidt.

Future attempts to correlate the clinical phenotypes with the biochemical ab-normalities and genomic lesions might be assisted if one could develop an agreed upon grading system for all patients with HGPRT deficiency. I would like to suggest two possibilities. First, a gading system based on the patient's overall functional ability as follows:

Grade 1: Totally dependent on others for the basic physical needs of daily living (e.g., toilet use, feeding, dressing) plus self-injurious behaviour.

Grade 2: As for grade 1 but without self-injurious behaviour.

Grade 3: Not totally dependent on others for their basic physical needs but requiring help with some aspects of daily living.

Grade 4: Able to lead an independent existence with only minor disabilities due to mild neurological involvement.

Grade 5: Able to lead a fully independent life without disability due to neuro-logical involvement and with the only disabilities being due to gout or urolithiasis.

An alternative grading based on the physical and biochemical findings is shown
in Table 1.

**Table 1.** HGPRT Deficiency: Patient grading based on physical and biochemical findings

| Finding | Grading | | | |
|---|---|---|---|---|
| | 1 | 2 | 3 | 4 |
| Hyperuricaemia (or hyperuricaciduria) | Yes | Yes | Yes | Yes |
| Organic neurological signs[a] | Yes | Yes | Yes | No |
| Self-injurious behaviour[b] | Yes | Yes | No | No |
| Mental handicap[c] | Yes | No | No | No |

[a] Any or all of: dystonic posturing, rigidity, hypotonia, dysarthria, long tract signs, nystagmus, incoordination.

[b] Self-injurious behavior at any time

[c] This is age related and requires careful psychometric testing in order to disentangle the extent of genuine mental developmental delay from the effects of environmental factors related to the extremely limited range of normal childhood activities in which these patients can participate.

The clinical phenotype associated with APRT deficiency is much more uniform.
There is less information about the range of genomic lesions in this disorder than
in (HGPRT) deficiency so that a similar grading scheme is not, in the present
state of knowledge, required.

# 2 The Clinical Aspects of HGPRT Deficiency

I. KAMILLI and U. GRESSER

The enzyme hypoxanthine-guanine phosphoribosyltransferase (HGPRT) catalyzes in the presence of 5-phosphoribosyl-1-pyrophosphate (PRPP) the formation of inosine-5-monophosphate and guanosine monophosphate from the bases hypoxanthine and guanine.

These reactions and similar ones are called "salvage pathways". The nucleotide metabolism depends on a orderly coordination of biosynthesis and salvage of bases. It is quite probable that in the human system only placenta and liver can satisfy their nucleotide demand by biosynthesis alone; all other organs, in varying degrees, depend on the salvage of bases.

Reduced or almost absent HGPRT activity will lead to a lack of nucleotides, and PRPP becomes increasingly available for the biosynthesis as a result of reduced consumption. Consequently purine biosynthesis and uric acid formation are increased and the patients suffer from hyperuricemia, gout and nephrolithiasis.

In addition, extensive or complete loss of HGPRT activity leads to a neurological phenomenon that was described for the first time by Catel and Schmidt in 1959 and became known as Lesch-Nyhan syndrome after the case reports by Lesch and Nyhan were published in 1964. Adult gout patients with heavily reduced HGPRT activity were described for the first time by Kelley et al. (1967).

## Inheritance

The HGPRT enzyme defect is transmitted X-chromosomally, which means that mothers carrying the defective gene do not contract the disease themselves (so-called carriers) but pass it on their sons. Since men, in contrast to women, only possess one set of X-chromosomally localized genes, men with this defective gene always fall ill. Therefore in the Anglo-American literature this condition is called "X-linked gout". Hara et al. described in 1982 the first case of a girl who showed typical clinical features and biochemical characteristics of the classical Lesch-Nyhan syndrome. Her mother was not a heterozygote for a deficiency of HGPRT. Possible genetic mechanisms were discussed.

# Clinical Symptoms

Depending on the level of HGPRT activity the clinical manifestations of HGPRT deficiency vary. A reduction of HGPRT activity to 5%–15% of the normal value leads to gout symptoms such as arthritis, nephrolithiasis, tophi and nephropathy. If HGPRT activity drops to 0%, in addition to gout a neurological condition develops with self-mutilation by biting, choreoathetosis, spasms and retarded mental development in association with growth retardation, the so-called Lesch-Nyhan syndrome.

## Complete HGPRT Deficiency

Patients with Lesch-Nyhan syndrome are clinically inconspicuous at birth. In isolated instances neurological symptoms may be observed during the first 3 months (Lesch and Nyhan 1964). At the age of 3–4 months, however, motor development is generally delayed. In addition, extrapyramidal symptoms in the form of discrete athetoid movement disorders of hands and feet, dystonia and chorea are observed between months 8 and 12 (Kelley and Wyngaarden 1983) (Fig. 1). These are at least in part responsible for the dysarthria that appears later on (Dreifuss et al. 1968). Signs of the pyramidal track lesions – hyperreflexia, clonus of the ankle joints, spasms – appear more or less after the first year and are responsible for the inability to walk in older children (Kelley and Wyngaarden 1983).

The typical symptom of Lesch-Nyhan syndrome, compulsive self-mutilation by biting, appears between the 2nd and 16th years (Fig. 2). Fingers, lips and cheeks are most commonly affected (Figs. 3, 4). This compulsive biting can develop to a degree where cuffs or bandages must be applied. In individual cases the only remedy is tooth extraction. In contrast to other neurological disorders associated with self-mutilation, this type of mutilation involving the loss of major parts of an organ seems to be characteristic of Lesch-Nyhan syndrome (Kelley and Wyngaarden 1983).

The patient's aggressiveness is directed not only at himself but also against relatives and strangers. Stress situations may intensify the aggressiveness and elicit opisthotonus. In about half of the patients convulsive attacks have been observed, that can not always be ascribed without any doubt to the enzymatic defect.

Technical examinations such as CSF diagnostics, electromyography, measurement of the velocity of nerve conduction and imaging or scanning procedures have all been without pathological findings. IQ in children with Lesch-Nyhan syndrome usually ranges from 39 to 65 (Kelley and Wyngaarden 1983).

A variety of transitional forms between Lesch-Nyhan syndrome and gout without neurological disorder have been described in patients with a partial enzymatic defect. The observations – individually or in variable combination – include retarded mental development, mild forms of quadriplegia, dysarthria,

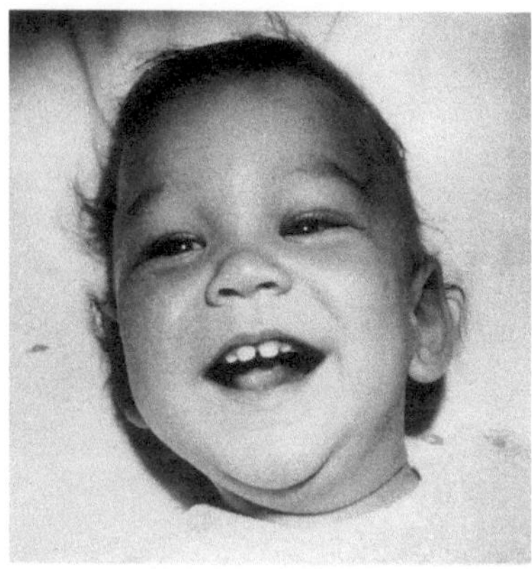

**Fig. 1.** A male patient with Lesch-Nyhan syndrome at the age of 1 year (no neurologic symptoms). (By kind permission of Y. S. Shin, Munich)

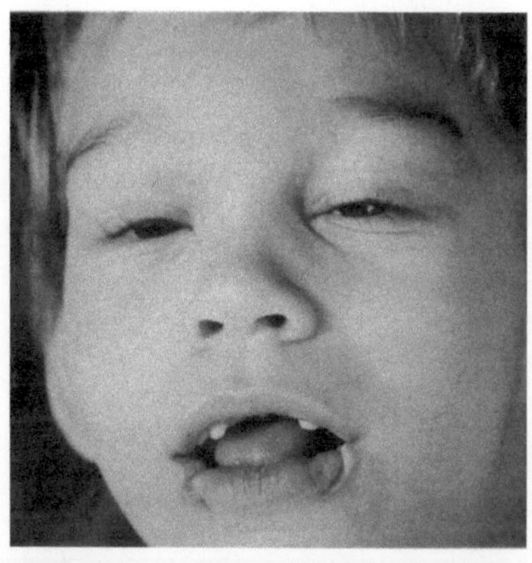

**Fig. 2.** The same boy at the age of 4 years. His lips are partially lost because of biting. (By kind permission of Y. S. Shin, Munich)

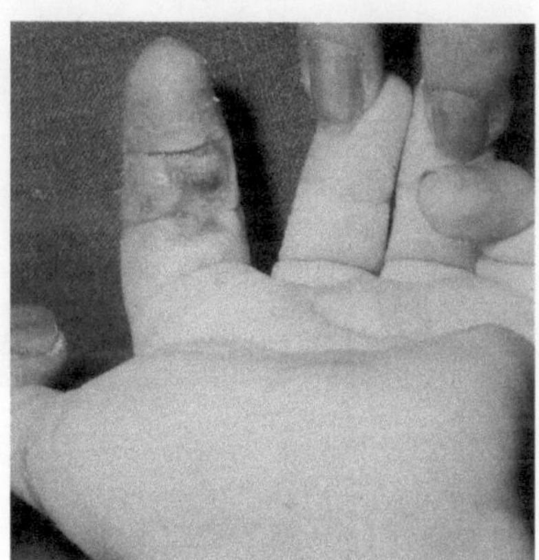

**Fig. 3.** The same boy: the fingers are hurt by biting (By kind permission of Y. S. Shin, Munich)

cerebellar ataxia and convulsive attacks. Patients do not show any tendency towards self-mutilation, however. In about 20% of all patients with partial HGPRT deficiency, neurological symptoms are found (Kelley et al. 1969; Kelley and Wyngaarden 1983). The connection between enzymatic defect and neurological disorder is unclear. Features common to all patients are the elevated uric acid values, frequently above 10 mg/100 ml, and the consequent manifestations of gout.

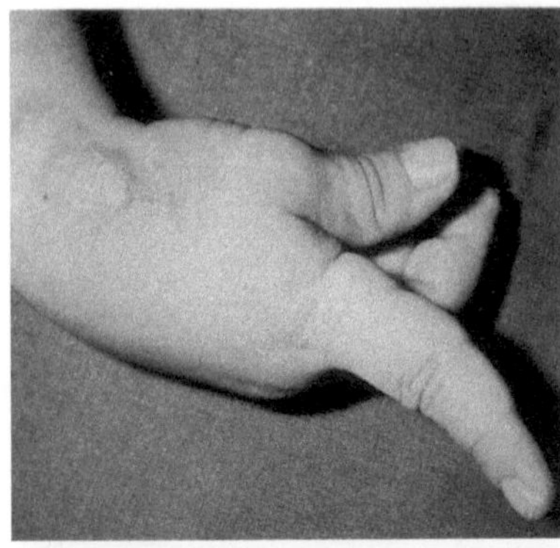

**Fig. 4.** The same boy: the typical athetoid movement of the fingers (By kind permission of Y. S. Shin, Munich)

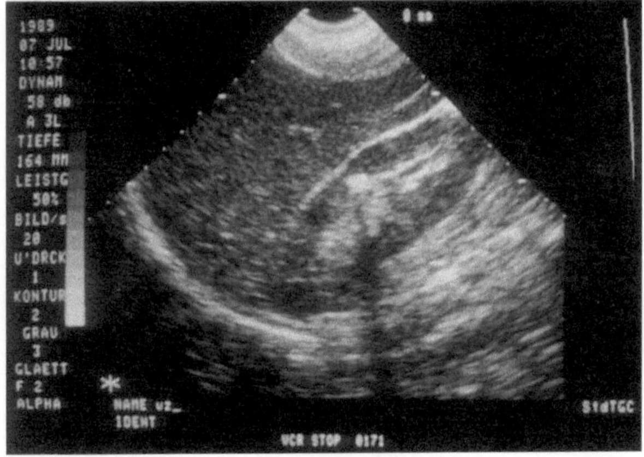

**Fig. 5.** Ultrasound of the right kidney of a 30-years-old patient with partial HGPRT deficiency: hyperechoic concrement in the renal pelvis

## Partial HGPRT Deficiency

The clinical consequence of partial HGPRT deficiency is hyperuricemia without neurological disorders. Some 75% of patients with partial HGPRT deficiency have a history of nephrolithiasis with multiple renal colics at a young age (Fig. 5). With chronic nephrolithiasis they have a danger of hydronephrosis followed by atrophic kidney. The elevated plasma uric acid values lead to uric acid deposits in the joints and thus to recurring acute attacks of gout (Fig. 6). If this condition exists for several years, tophus formation is possible (Fig. 7).

*Case report.*   A patient from Yugoslavia, now 60 years old, has been in the care of our medical outpatient department for 28 years (Zöllner et al. 1978; Kamilli et al. 1993). In 1964 he came to our hospital for the first time with fistulating

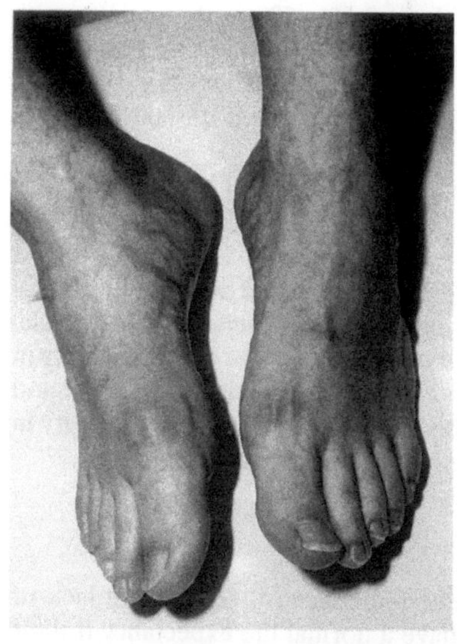

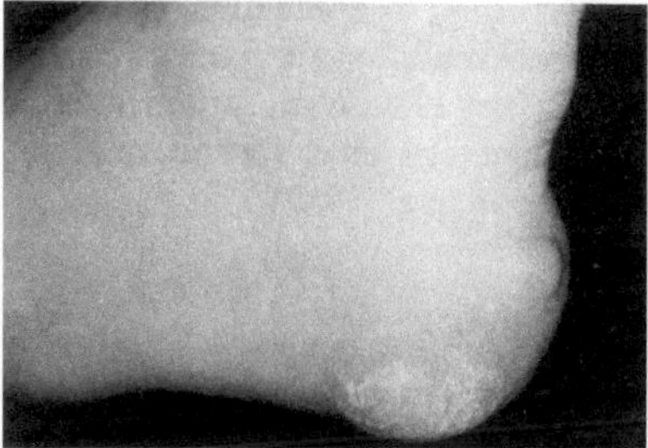

**Fig. 7.** Tophus in the heel of an 28-year-old patient with incomplete HGPRT deficiency and fistulating tophic gout since the age of 18

**Fig. 6.** Acute attack of gout in the first metatarsophalangeal joint of a 42-year-old patient with partial HGPRT deficiency

tophic gout and poor general health. He reported that he had suffered from multiple gout attacks in almost all joints since the age of 18, while nephrolithiasis with multiple colics had been discovered when he was 26. At admission to our hospital the plasma uric acid level was 12 mg/100 ml and uric acid excretion 1g/day. The neurological state was normal. Under consistent therapy with allopurinol 300 mg/day no more gout attacks and renal colics occurred and the tophi slowly disappeared. The patient still takes allopurinol, and with a plasma uric acid level of 5.5 mg/100 ml his quality of life is not impaired.

## Other Symptoms, Epidemiology

Patients with HGPRT deficiency frequently display a macrocytic blood count and megaloblastic alterations of bone marrow with and without anemia. A relative deficiency of folic acid was assumed after low folic acid levels had been measured in some patients (Kelley et al. 1969); folic acid is needed for purine biosynthesis. The connections, however, are not conclusive in vivo or in cell cultures. In one patient it was possible to normalize the blood count by the administration of adenine, while folic acid substitution was ineffective (van der Zee et al. 1970).

Immunological defects, earlier assumed to be present in HGPRT deficiency, have not been established. Most patients with Lesch-Nyhan syndrome die of renal failure and infections in early adulthood. The commonly occurring pneumonias and urinary tract infections have been attributed to aspiration or obstruction of the urinary tract by nephrolithiasis.

According to epidemiological studies about 0.5%–5% of the patients with hyperuricemia have a partial HGPRT deficiency (Davidson et al. 1989; Stout and Caskey 1989; Wyngaarden and Kelley 1983). The prevalence of Lesch-Nyhan syndrome (complete absence of HGPRT) is estimated at 1 in 100 000 live births (Davidson et al. 1989).

## Diagnosis

If HGPRT deficiency is suspected the HGPRT activity in red blood cells or fibroblasts is assayed by high-pressure liquid chromatography. There is a good correlation between the degree of the neurological symptoms and the enzyme activity in intact fibroblasts (Page et al. 1981), but not in the blood cells (Emmerson and Thompson 1973; Rijksen et al. 1981). In the case of minimal or absent activity in red blood cells one must not necessarily expect neurological symptoms.

## Therapy

It can be stated generally that, in contrast patients with a complete lack of HGPRT, those with a partial deficiency have a normal life expectancy if their hyperuricemia is treated adequately. Dietary purine restriction, high fluid intake and allopurinol are the therapy of choice.

The high effectiveness of allopurinol, however, involves the danger of a complication that has observed only in patients with increased uric acid formation. There is increased oxipurine excretion attributed to an increase in the excretion of xanthine, rather than hypoxanthine, so xanthine calculi or acute xanthine nephropathy may appear (Klinenberg et al. 1965). As a consequence the allopurinol dose should be kept as low as possible.

However, there is no effective therapy for the neurological complications of Lesch-Nyhan syndrome. The ultimate goal is the management of this inborn error of metabolism at the genetic, rather than the symptomatic level.

## Conclusions

Partial HGPRT deficiency should be considered in any young patient with hyperuricemia and nephrolithiasis and/or gout, and the activity of the enzyme should be determined.

## References

Catel W, Schmidt J (1959) Über familiäre gichtische Diathese in Verbindung mit zerebralen und renalen Symptomen bei einem Kleinkind. Dtsch Med Wochenschr 84: 2145–2148

Davidson BL, Palella TD, Kelley WN (1989) Hypoxanthine-guanine-phosphoribosyl-transferase deficiency: Molecular basis and clinical relevance. In: Wolfram G (ed), Genetic and therapeutic aspects of lipid and purine metabolism. Springer, Berlin Heidelberg New York, pp 57–66

Dreifuss FE, Newcombe DS, Shapiro SL, Sheppard GL (1968) X-linked primary hyperuricemia (hypoxanthine-guanine-phosphoribosyltransferase deficiency encephalopathy). J Ment Defic Res 12: 100–105

Emmerson BT, Thompson L (1973) The spectrum of hypoxanthine-guanine phosphoribosyltransferase deficiency. Q J Med 42: 423–440

Hara K, Kashiwamata S, Ogasawara N, Ohishi H, Natsume R, Tsutomu Y, Hakamada S, Miyazaki S, Watanabe K (1982) A female case of the Lesch-Nyhan syndrome. Tohoku J Exp Med 137: 275–282

Kamilli I, Gresser U, Gathof B, Gröbner W (1993) Partial HGPRT-deficiency, pheochromocytoma and erythrocytosis. J Inherit Metab Dis (in print)

Kelley WN, Rosenbloom FM, Henderson JF, Seegmiller JE (1967) A specific enzyme defect in gout associated with overproduction of uric acid. Proc Natl Acad Sci USA 57: 1735–1739

Kelley WN, Greene ML, Rosenbloom FM, Henderson JF; Seegmiller JE (1969) Hypoxanthine-guanine phosphoribosyltransferase deficiency in gout. Ann Intern Med 70: 155–206

Kelley WN, Wyngaarden JB (1983) Clinical symptoms associated with hypoxanthine-guanine phosphoribosyltransferase deficiency. In: Stanbury JB; Wyngaarden JB, Fredrickson DS, Goldstein JL, Brown MS (eds) The metabolic basis of inherited disease, 5th edn. McGraw-Hill, New York, pp 1115–1143

Klinenberg JR; Goldfinger SE, Seegmiller JE (1965) The effectiveness of the xanthine oxidase inhibitor allopurinol in the treatment of gout. Ann Intern Med 62: 639–647

Lesch M, Nyhan WL (1964) A familial disorder of uric acid metabolism and central nervous system function. Am J Med 36: 561–570

Page T, Baky B, Nissinen E, Nyhan WL (1981) Hypoxanthine-guanine phosphoribosyltransferase variants: Correlation of clinical phenotype with enzyme activity. J Inherit Metab Dis 4: 203–206

Rijksen G, Staal GEJ, van der Vlist MJM (1981) Partial hypoxanthine-guanine phosphoribosyltransferase deficiency with full expression of the Lesch-Nyhan syndrome. Hum Genet 57: 39–47

Stout T, Caskey T (1989) Hypoxanthine-phosphoribosyltransferase deficiency: the Lesch-Nyhan syndrome and gouty arthritis. In: Scriver CR, Beaudet AL, Sly WS, Valle D (eds) The metabolic basis of inherited disease, 6th edn. McGraw-Hill, New York 1007–1028

Wyngaarden JB, Kelley WN (1983) Gout. In: Stanbury JB, Wyngaarden JB, Fredrickson DS, Goldstein JL, Brown MS (eds) The metabolic basis of inherited disease, 5th edn. McGraw-Hill, New York, pp 1043–1114

Van der Zee SPM, Lommen EJP, Trijbels JMF, Schretlen ED (1970) The influence of adenine on the clinical features and purine metabolism in the Lesch-Nyhan syndrome. Acta Paediatr Scand 59: 259–264

Zöllner N, Goebel FD, Gröbner W (1978) Partial HGPRT deficiency: persistence of tophi after 12 years of therapeutic normouricemia and development of a pheochromocytoma. Monogr Hum Genet 10: 112–115

# 3 The Biochemical Basis of HGPRT Deficiency

J. G. PUIG and F. A. MATEOS

## Introduction

Hypoxanthine guanine phosphoribosyltransferase (E.C.2.4.2.8.; HGPRT) deficiency is associated with a marked clinical heterogeneity [1–3]. All patients with HGPRT deficiency exhibit a common characteristic: increased uric acid production [3]. However, the clinical consequences of this uric acid overproduction may differ markedly from patient to patient. This variability depends not only on the rate of uric acid production but also on the factors that determine urate precipitation, some of which remain to be fully delineated. Among the known factors associated with urate precipitation, effective hypouricemic treatment has dramatically changed the spectrum of the clinical manifestations related to urate deposition in HGPRT deficient patients. In addition, HGPRT deficiency may be associated with a neurological syndrome that in its full expression is characterized by spasticity, hyperreflexia, choreoathetoid movements, mental retardation, and compulsive self-injurious behavior (Fig. 1) [4–6].

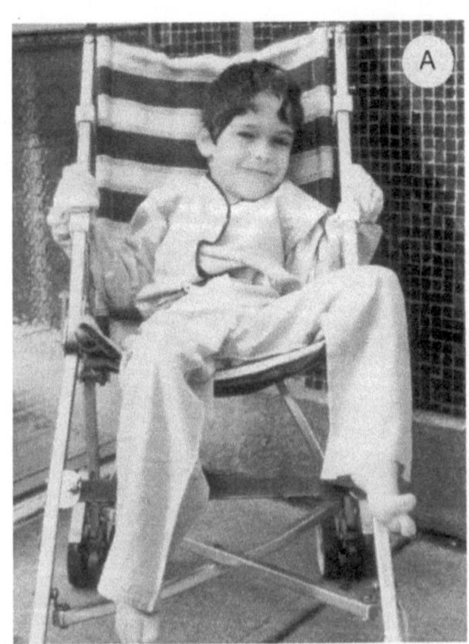

**Fig. 1 A–C.** A 5-year-old boy with Lesch-Nyhan syndrome. **A** Spontaneous Babinski sign of his left foot is depicted. **B** and **C** self-injury compulsive behavior when the hand protection is removed.

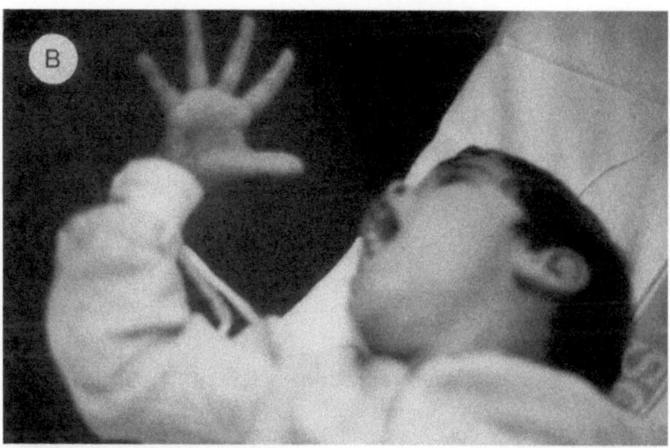

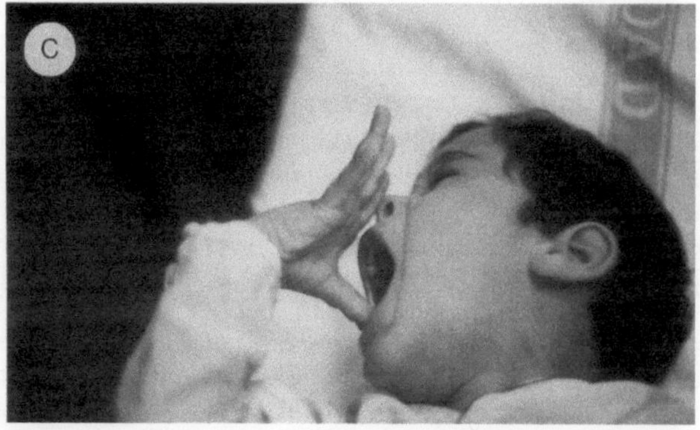

**Fig. 1B, C**

In contrast, some patients with HGPRT deficiency do not exhibit these neurological symptoms and others manifest different degrees of spasticity, dystonia, hyperreflexia, or mental retardation ($\sim 20\%$ of the patients with partial HGPRT deficiency) [7]. For comprehensive purposes, patients with HGPRT deficiency have been classified into those who self-mutilate (Lesch-Nyhan syndrome, "complete HGPRT deficiency") and those who do not exhibit a self-injurious behavior ("partial HGPRT deficiency", Kelley-Seegmiller syndrome).

Since its initial description, HGPRT deficiency has been the subject of extensive investigations that contributed remarkably to our understanding of purine metabolism [8]. Studies have focused on three fundamental areas: (a) the molecular basis of the primary defect in HGPRT activity, (b) the biochemical mechanisms responsible for the overproduction of uric acid, and (c) the pathophysiology of the neurological abnormalities associated with HGPRT deficiency. The precise molecular basis of HGPRT deficiency is reviewed elsewhere in this volume. In this article we will focus on the progress made in clarifying the last two areas.

# Biochemical Mechanisms of Excessive Uric Acid Production

The excessive synthesis of uric acid in patients with HGPRT deficiency has been related to the enzyme defect. The most direct evidence for this relation in vivo would be provided by documenting that inhibition of HGPRT activity is accompanied by an increase in uric acid synthesis and by demonstrating that uric acid overproduction in HGPRT deficient patients is normalized by enzyme replacement therapy. However, an accelerated rate of purine biosynthesis de novo has been demonstrated in human lymphoblasts [9] and in mouse neuroblastoma cells selected for a deficiency of HGPRT following mutagenesis.

The increased production of uric acid in HGPRT deficient patients has been demonstrated by several techniques. The most striking biochemical abnormality in the first patients reported with the Lesch-Nyhan syndrome and partial HGPRT deficiency was an increased serum urate concentration and a markedly elevated urinary uric acid excretion [4, 10]. Moreover, using radioactive purine precursors, such as [$^{14}$C] or [$^{15}$N]glycine, it was evident that these patients incorporated these precursors into uric acid at an accelerated rate and that the cumulative urinary radioactive excretion was markedly elevated [4, 7, 11]. In addition, the administration of tracer doses of [$^{8-14}$C]adenine to HGPRT deficient patients was followed by an increased cumulative radioactive excretion [12, 13] (Fig. 2). By a sensitive and specific high-pressure liquid chromatography (HPLC) procedure, we documented increased baseline plasma concentrations and urinary excretion of hypoxanthine and xanthine in HGPRT deficient patients as compared to normal subjects and patients with primary gout [14]. These preliminary results have been extended to include ten patients with HGPRT deficiency (three with partial HGPRT deficiency and seven with the Lesch-Nyhan syndrome). Plasma and 24 h urinary uric acid, hypoxanthine, and xanthine concentrations were markedly increased in the enzyme deficient patients as compared to 45 patients with primary gout and 25 normal subjects (Table 1). To examine purine nucleotide degradation in HGPRT deficiency, four patients received an intravenous infusion of fructose (0.5 g/kg of body weight in 15 min). Fructose infusion elicited a dramatic increase in total urinary purines (sum of hypoxanthine, xanthine, and uric acid) in HGPRT deficient patients as compared to normal subjects [13] (Fig. 3). Taken together these data indicate that purine overproduction in HGPRT deficiency is due to both an increased purine synthesis de novo and an enhanced purine nucleotide degradation.

Two possible mechanisms have been proposed to explain the excessive purine synthesis in HGPRT deficiency (Fig. 4). First, the decreased reutilization of purine bases for ribonucleotide synthesis (IMP, GMP) may lead to a decreased intracellular concentration of these nucleotides which normally exert an inhibitory effect on de novo purine biosynthesis [15, 16]. The lack of the HGPRT-mediated inhibitory effect on de novo purine synthesis prompts an accelerated rate of this pathway as a compensatory mechanism for purine nucleotide homeostasis. This theory may explain the finding of an increased de novo IMP synthesis in HGPRT deficient fibroblasts to compensate for the lack of IMP

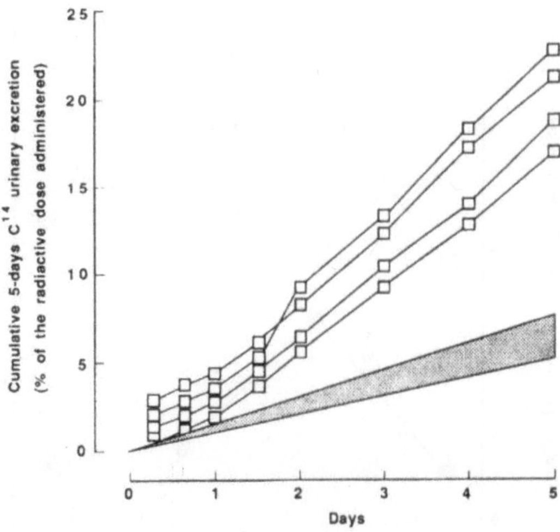


**Fig. 2.** Cumulative urinary radioactivity after intravenous infusion of tracer doses of [8–$^{14}$C]adenine. Five days after the infusion of [8–$^{14}$C]adenine, four normal subjects excreted (mean ± SEM) 6.0% ± 0.4% of the administered radioactive dose (*shaded area*). Four patients with HGPRT deficiency (□) excreted 19.4% ± 1.7% of the administered radioactivity

**Table 1.** Plasma and 24h urinary concentrations of purines in normal subjects, patients with primary gout, and HGPRT deficient patients

| | Normal subjects ($n = 25$) | Primary gout ($n = 45$) | HGPRT deficiency ($n = 10$) |
|---|---|---|---|
| **Plasma** | | | |
| Urate (µmol/l) | 292 ± 60 | 446 ± 83[a] | 568 ± 149[b] |
| Hypoxanthine (µmol/l) | 2.3 ± 1.1 | 4.4 ± 2.1[a] | 6.6 ± 2.0[b] |
| Xanthine (µmol/l) | 0.7 ± 0.2 | 1.2 ± 0.6[a] | 2.1 ± 0.6[b] |
| **Urine** | | | |
| Uric acid (mmol/24h/1.73 m$^2$) | 2.63 ± 0.51 | 2.52 ± 0.92 | 7.41 ± 3.85[b] |
| Hypoxanthine (µmol/g creatinine) | 33 ± 9 | 21 ± 10[a] | 550 ± 375[b] |
| Xanthine (µmol/g creatinine) | 22 ± 9 | 15 ± 9[a] | 227 ± 176[b] |

[a] $p < 0.01$ vs normal subjects.
[b] $p < 0.01$ vs normal subjects and patients with primary gout.

synthesis from hypoxanthine reutilization [17]. In addition, it explains the observed increase in de novo purine synthesis following incubation of hepatocytes with fructose, which elicits an intracellular depletion of adenine nucleotides [18]. The hypothesis that in HGPRT deficiency purine biosynthesis de novo is increased to compensate for the diminished formation of purine nucleotides from reutilization of purine bases is supported by the finding that HGPRT deficient cells with an intact capacity for purine synthesis de novo exhibit normal

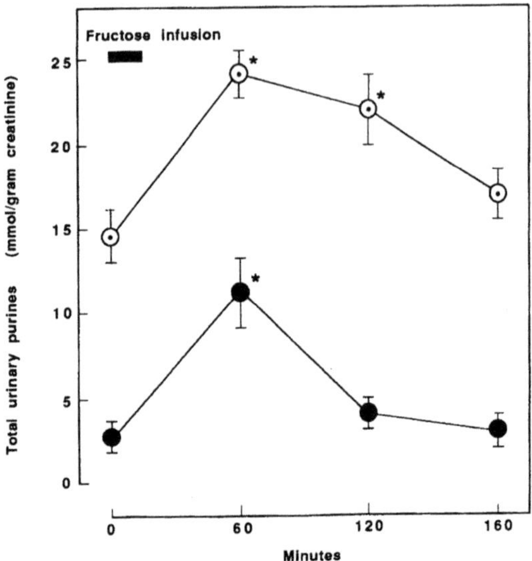



**Fig. 3.** Total urinary purine excretion in response to intravenous fructose infusion in normal subjects and in HGPRT deficient patients. The absolute increase in total urinary purines (sum of uric acid, hypoxanthine, xanthine and inosine) in four HGPRT deficient patients (○) was (mean ± SEM) 17.96 ± 3.36 and 10.38 ± 3.80 mmol/g creatinine in four normal subjects (●) ($p < 0.05$). *Asterisk* significance at $p < 0.01$ level with respect to baseline values. *Dot inside the circle*, significance at $p < 0.01$ level with respect to normal subjects. (Adapted from [13])

adenine nucleotide concentrations [19, 20], whereas Lesch-Nyhan patients cells with a limited capacity for de novo purine synthesis show a diminished concentration of adenine nucleotides [21, 22]. In four patients with HGPRT deficiency we examined whether the increased purine biosynthesis de novo contributes to adenine nucleotide formation compared with normal subjects. Four HGPRT deficient patients (three with the Lesch-Nyhan syndrome and one with partial HGPRT deficiency) and four normal subjects received tracer doses of [8-$^{14}$C]-adenine to radiolabel the adenine nucleotide pool and 5 days later a rapid infusion of fructose to stimulate purine nucleotide degradation. Fructose infusion increased urinary radioactivity in normal subjects to (mean ± SEM) 106% ± 102% of the baseline values and to 141% ± 13% in the enzyme deficient patients ($p<0.01$). The absolute mean increase in total urinary purines in controls was 10.38 ± 3.80 and 17.96 ± 3.36 mmol/g creatinine in the patients ($p < 0.05$). The apparent specific radioactivity of urinary purines decreased in the enzyme deficient patients from a mean of $1.66 \cdot 10^5$ to $1.38 \cdot 10^5$ cpm/mmol of purines but increased in control subfects from a mean of $1.29 \cdot 105$ to $3.64 \cdot 10^5$ cpm/mmol ($p < 0.02$) (Fig. 5A). To assess whether the decrease in the specific activity of urinary purines was due to an elevated rate of nonradiolabeled de novo purine synthesis, two HGPRT deficient patients were treated with allopurinol and adenine followed 5 days later by a fructose infusion. The administration of adenine increased the specific activity of urinary purines after the infusion of fructose from a mean baseline value of $1.05 \cdot 10^5$ to $1.42 \cdot 10^5$ cpm/mmol of purines and this effect was probably due to inhibition of de novo purine biosynthesis (Fig. 5B). This shift in urinary purine-specific activity in both patients (from a mean decrease of 19% on no drugs to a mean increase of 35% on adenine therapy) could not be accounted for by the simultaneous allopurinol treatment which was given to prevent adenine toxicity. From this study we concluded that, when the salvage pathway for hypoxanthine reutilization is impaired, purine synthesis de novo makes an enhanced contribution to adenine nucleotide turnover (Fig. 4) [14].

It has been calculated that there is a 20-fold acceleration in the rate of de novo IMP synthesis in HGPRT deficiency [15, 17, 23]. This accelerated rate may be crucial to maintain normal adenine nucleotide concentrations [17, 19, 20]. This is consistent with the observation that de novo IMP synthesis in HGPRT deficient fibroblasts is mainly channeled to adenine nucleotide synthesis [17]. However, certain cells may be particularly dependent on purine reutilization for nucleotide synthesis and it is possible that de novo purine synthesis may not make a sufficient contribution to maintain adequate nucleotide homeostasis.

The second mechanism by which HGPRT deficiency prompts an enhanced purine synthesis de novo is due to an increased concentration of phosphoribosyl pyrophosphate (PRPP). This compound is a common substrate for amidophos-

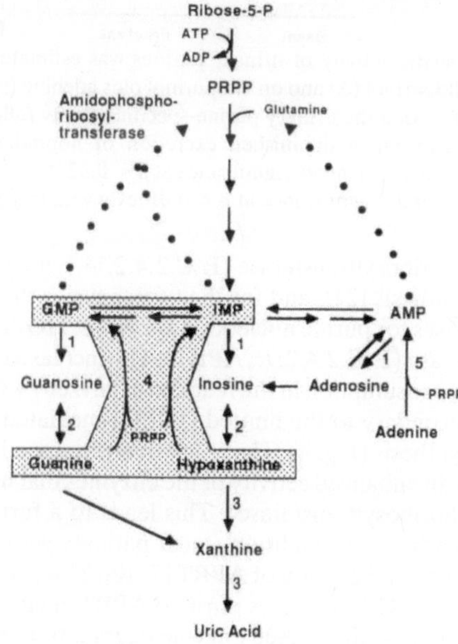


**Fig. 4.** Purine nucleotide synthesis and catabolism. The intracellular nucleotide pool is maintained by interdependent metabolic processes. Purine biosynthesis de novo is a pathway for the synthesis of the purine ring from nonpurine precursors. The availability of phosphoribosyl pyrophosphate (*PRPP*) is ratelimiting for de novo purine synthesis. Purine nucleotide catabolism poceeds through 5'-nucleotidase (*1*), purine nucleoside phosphorylase (*2*), and xanthine oxidase (*3*). In humans the end product of purine nucleotide degradation is uric acid. The reutilization of purines is catalized by hypoxanthine guanine phosphoribosyltransferase (HGPRT) (*4*) and adenine phosphoribosyltransferase (*5*). These pathways allow the conversion of purine bases to their respective nucleotides (*IMP, GMP,* and *AMP*). A deficiency of HGPRT (*4*) leads to an inability to reutilize hypoxanthine and guanine (*shaded area*), and the total amount of these compounds formed is oxidized to uric acid. Purine synthesis de novo is enhanced to compensate for the lack of nucleotide synthesis from purine reutilization. In addition, PRPP is not consumed in the reaction catalized by HGPRT and its increased availability enhances de novo purine synthesis

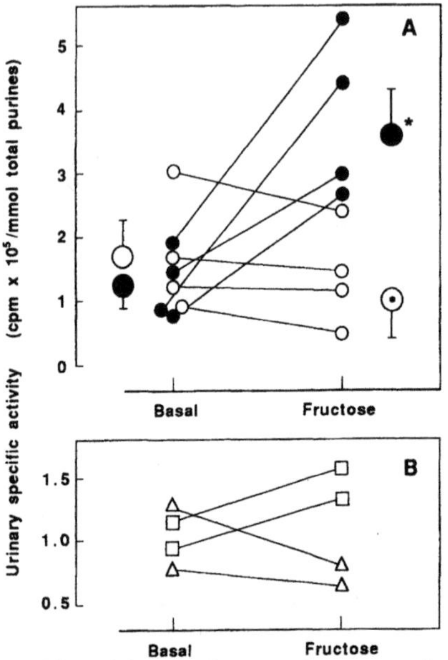

**Fig. 5 A,B.** Apparent specific activity of urinary purines before and after the infusion of fructose in normal subjects and in patients with HGPRT deficiency. **A** The urinary purine-specific activity was estimated for 1 h before (*basal*) and 1 h after the intravenous infusion of fructose in four normal subjects (●) and in four patients with HGPRT deficiency (○). *Error bars* represent SEM. The specific activity of urinary purines increased in controls from a baseline value of $1.29 \pm 0.25 \cdot 10^5$ to $3.64 \pm 0.72 \cdot 10^5$ cpm/mmol total purines but decreased in HGPRT deficient patients from a baseline level of $1.66 \pm 0.81 \cdot 10^5$ to $1.38 \pm 0.65 \cdot 10^5$ cpm/mmol purines, probably due to an increased excretion of nonradiolabeled purines from de novo purine synthesis which is enhanced by the consumption of adenine nucleotides to metabolize fructose. **B** In two enzyme deficient patients the specific activity of urinary purines was estimated before and after fructose infusion on allopurinol (△) and on allopurinol plus adenine (□) therapy. The administration of adenine increased the urinary purine-specific activity following fructose infusion in both patients, suggesting a diminished excretion of nonradiolabeled purines from de novo purine synthesis. *Asterisk* significance at $p < 0.02$ level with respect to baseline values. *Dot inside the circle*, significance at $p < 0.01$ level with respect to normal subjects

phoribosyltransferase (E.C.2.4.2.14.), a rate limiting enzyme in de novo purine synthesis [24], and for the two enzymes that catalyze the reutilization of purine bases for purine nucleotide synthesis: HGPRT and adenine phosphoribosyltransferase (E.C.2.4.2.7.; APRT). The increased availability of PRPP is due to its lack of consumption in the reaction catalyzed by HGPRT and to its increased synthesis secondary to the limited HGPRT-mediated inhibitory effect on de novo purine synthesis (Fig. 4). The result of an increased intracellular concentration of PRPP is an enhanced activity of the enzymes that utilize this substrate (i.e., amidophosphoribosyltransferase). This leads to a further increase in de novo purine biosynthesis. In addition, most patients with HGPRT deficiency also exhibit an increased activity of APRT [7, 10]. This finding has been explained on the basis that PRPP stabilizes purified APRT in vitro [25] and that this stabilization may lead to a diminished enzyme degradation rate and thus to increased APRT activity [25]. This hypothesis may explain why both APRT activity and the enzyme protein were more stable in erythrocytes from Lesch-Nyhan patients than in those from normal subjects [26] and that a direct relation was found between intracellular PRPP concentrations and erythrocyte APRT activity [27]. In fact, some patients with partial HGPRT deficiency have been described who showed normal erythrocyte APRT activity and PRPP concentrations [28].

The accelerated purine synthesis de novo and enhanced purine nucleotide degradation, characteristic of patients with HGPRT deficiency, result in uric acid overproduction, but this is not a uniform phenomenon in all tissues. The enzyme responsible for uric acid synthesis (xanthine oxidase, E.C.1.2.3.2.) is predominantly located in the liver, the small intestine [29], and in the endothelial cells of capillaries [30]. Thus, overproduction of uric acid is probably restricted to those tissues in which the increased concentrations of hypoxanthine and xanthine are finally oxidized to uric acid.

## Pathogenesis of the Neurological Dysfunction

The mechanism by which a decrease in HGPRT activity causes the dramatic neurological symptoms observed in patients with Lesch-Nyhan syndrome and in some patients with partial HGPRT deficiency remains to be fully delineated [31]. Anatomical studies have shown that the brain is slightly small for the age of the patients with Lesch-Nyhan syndrome, but no gross structural or microscopic abnormalities have been found [6, 32–34]. In particular, no cortical atrophy or reduction in white matter is evident and the basal ganglia and substantia nigra show no gliosis or neuronal depletion. In developing rat brain HGPRT activity progressively increases with age and PRPP amidotransferase progressively decreases [35]. The finding that in normal persons the activity of HGPRT is highest in the brain and within the brain it is highest in the basal ganglia [7] suggests that these areas are particularly dependent on the salvage pathway for purine nucleotide synthesis. In patients with the Lesch-Nyhan syndrome, HGPRT activity in the brain has been found to be markedly reduced [6, 36, 37]. The neurological dysfunction has been related to the enzyme defect, being more severe in Lesch-Nyhan patients than in subjects with residual HGPRT activity [1–3]. However, one patient has been reported with a moderately diminished erythrocyte HGPRT activity that showed the characteristics of the Lesch-Nyhan syndrome [38], and patients with a lower HGPRT activity than that usually found in Lesch-Nyhan patients did not manifest neurological symptoms [2]. Unfortunately, a noninvasive procedure to determine HGPRT activity in the central nervous system (CNS) of living subjects is not available and an inverse relationship between the severity of the neurological symptoms and the CNS HGPRT activity has not been established. However, a significant inverse relationship has been found between HGPRT activity in intact fibroblasts and the severity of the neurological symptoms [39].

What are the potential pathomechanisms underlying the neurological disorder in all patients with complete HGPRT deficiency and in some patients with partial HGPRT deficiency? It is possible that this enzyme deficiency may cause a derangement in the concentration of certain substances necessary for normal CNS function and/or that a toxic product accumulates as a result of HGPRT deficiency. Uric acid, hypoxanthine, and xanthine are potential toxic substances for the CNS. However, in the human brain xanthine oxidase is virtually absent [40] and hypoxanthine and xanthine are the physiological end products of purine nucleotide degradation. In patients with HGPRT deficiency, urate concentrations in the cerebrospinal fluid (CSF) are normal [36, 41] and urate deposits in the CNS have not been documented 6, 42]. By means of a spectrophotometric method Kelley et al. [7] documented increased CSF oxypurine (hypoxanthine plus xanthine) concentrations in four patients with HGPRT deficiency. Plasma oxypurines were normal and the mean CSF/plasma oxypurine ratio was three fold elevated as compared to normal subjects [7]. Using an HPLC method we simultaneously determined plasma and CSF inosine, hypoxanthine, xanthine, and uric acid concentrations in four patients with HGPRT deficiency (three with the Lesch-Nyhan syndrome and one with partial HGPRT deficiency) and in four age-matched controls (Table 2) [43]. Plasma and CSF hypoxanthine and xanthine concentrations were markedly increased in the enzyme deficient patients. Moreover, mean CSF hypoxanthine concentration was increased five fold whereas mean xanthine levels were increased two fold. As hypoxanthine and xanthine in the CNS are basically the end products of adenine and guanine nucleotides, respectively, the predominance of hypoxanthine over xanthine may suggest that in the CNS of these patients adenine nucleotide degradation is more pronounced than guanine nucleotide catabolism. To assess the transfer of purines through the hematoencephalic barrier, plasma and CSF purines were determined following allopurinol (5–10 mg/kg of body weight per day) administration for 7 days. Allopurinol was not expected to markedly increase

**Table 2.** Plasma and cerebrospinal fluid purine concentrations in HGPRT deficient patients in the basal state and following allopurinol administration[a]

|  | Normal subjects | HGPRT deficiency ($n = 4$) | |
|  | ($n = 4$) | Basal state | Allopurinol |
|---|---|---|---|
| **Plasma** | | | |
| Urate (μmol/l) | 286 ± 12 | 593 ± 114[b] | 205 ± 66[c] |
| Hypoxanthine (μmol/l) | 1.7 ± 0.4 | 8.7 ± 1.6[b] | 38.6 ± 1.9[c] |
| Xanthine (μmol/l) | 0.9 ± 0.2 | 2.0 ± 0.3[b] | 21.0 ± 3.5[c] |
| Inosine (μmol/l) | 0.9 ± 0.2 | 1.6 ± 0.5[b] | 5.1 ± 1.0[c] |
| **Cerebrospinal Fluid** | | | |
| Urate (μmol/l) | 12 ± 9 | 26 ± 6 | ND |
| Hypoxanthine (μmol/l) | 3.3 ± 1.1 | 17.5 ± 2.8[b] | 35.0 ± 4.6[c] |
| Xanthine (μmol/l) | 2.0 ± 0.2 | 4.2 ± 0.3[b] | 11.9 ± 1.9[c] |
| Inosine (μmol/l) | 0.6 ± 0.2 | 0.6 ± 0.1 | 0.6 ± 0.1 |

[a] Allopurinol (5–10 mg/kg of body weight) was administered for 7 days. Data are means ± SEM; ND, not detectable.
[b] $p < 0.01$ for the comparison with normal subjects.
[c] $p < 0.01$ for the comparison with the basal state.

CSF oxypurine levels since xanthine oxidase is virtually absent in the CNS [40]. The administration of allopurinol substantially increased plasma inosine, hypoxanthine, and xanthine concentrations in the patients. Simultaneously, CSF hypoxanthine increased three fold and CSF xanthine two fold with respect to baseline values (Table 2) [43]. These results indicate that in HGPRT deficiency plasma oxypurines may be transferred to the CSF. However, the similar CSF hypoxanthine and xanthine concentrations in our patient with partial HGPRT deficiency and in our three patients with Lesch-Nyhan syndrome, and the failure of allopurinol treatment to modify the neurological symptoms indicate that oxypurines are not in themselves responsible for the CNS dysfunction associated with HGPRT deficiency.

The observations that HGPRT activity is normally higher in the CNS than in other tissues and that the activity of PRPP amidotransferase is low [44] suggest that the synthesis of IMP and GMP may be particularly dependent on the reutilization of hypoxanthine and guanine, respectively. In the absence of HGPRT activity, purine nucleotides may be synthesized entirely de novo. An increased activity of this pathway may eventually lead to depletion of certain essential cofactors, such as folic acid or ATP, for normals CNS development or function. However, intracellular folate and ATP levels are normal in Lesch-Nyhan patients [45]. There remains the possibility that the limited brain purine biosynthesis de novo, in the absence of purine reutilization, could lead to depletion of certain nucleotide pools (i.e., guanine nucleotides). Several observations have suggested that depletion of guanine nucleotides could explain the neurological symptoms of Lesch-Nyhan patients. Guanine nucleotides modulate dopaminergic receptors [46] and depletion of guanine nucleotides may cause a regulation receptor disturbance with dopaminergic hypersensitivity [47]. This hypothesis is supported by the following findings: (a) the brains of three patients with the Lesch-Nyhan syndrome showed a 10%–30% decrease in the function of dopamine neuron terminals [37]; (b) the administration of apomorphine (a dopamine agonist) to rats rendered supersensitive to dopamine by unilateral intracerebral injection of 6-hydroxydopamine into the nigrostriatal area caused an incessant biting of the limbs and tail [48]; (c) self-mutilation was elicited in monkeys by the administration of apomorphine or levodopa plus a dopadecarboxylase inhibitor after unilateral surgical lesions in the ventromedial tegmental area of the brainstem [49, 50]. This behavior was stopped when the drugs were withdrawn or when the dopamine receptor antagonist fluphenazine was given [49, 50]; (d) self-injury abated in a Lesch-Nyhan patient who was treated with fluphenazine [51], although in another patient, reported on by Goldstein et al. [51], and in two additional patients with the Lesch-Nyhan syndrome treated by us with fluphenazine self-mutilating behavior was not essentially modified. The hypothesis of guanine nucleotide depletion in HGPRT deficiency has not been fully documented. Erythrocyte guanine nucleotide levels have been reported to be slightly diminished in Lesch-Nyhan patients [45, 52], but the importance of this findings is limited by the fact that erythrocytes do not have the ability to synthesize purines de novo. We have provided additional evidence favoring in vivo GTP depletion in HGPRT deficiency [13]. By means of a rapid intravenous fructose infusion we sought to determine that there were differences in the rate of GTP degradation in four HGPRT deficient patients as compared to four normal subjects. The normal plasma guanosine response following fructose infusion was an increase from undetectable levels to $1.5 \pm 0.6\,\mu M$ (Fig. 6.). Patients with HGPRT deficiency showed no increase of plasma guanosine levels in response to the infusion of fructose. In addition, when the four HGPRT deficient patients were treated with allopurinol, fructose infusion elicited a lower increase in plasma and urinary xanthine concentrations than in two patients with hereditary xanthinuria [53]. The hypothesis that in HGPRT deficiency GTP depletion causes dopaminergic hypersensitivity and that this may explain the neurological symptoms is obscured by the fact that GTP depletion has also been reported in purine nucleoside phosphorilase (PNP) deficiency and PRPP synthetase overactivity [45, 52, 54]; these enzyme defects are not commonly associated with neurological impairment.

Additional pathological findings in HGPRT deficient patients that may be related to (or be consequence of) an impaired neronal function include:

1. Decreased plasma dopamine-β-hydroxylase activity with a reduced increase of plasma norepinephrine to emotional or postural stress as compared to normal subjects [55, 56]. However, hypothalamic levels of dopamine-β-hydroxylase have been found to be normal [37], suggesting that certain biochemical observations made in plasma or CSF from patients with the Lesch-Nyhan syndrome may not be representative of CNS function.
2. Serotonin dysfunction suggested by the finding of reduced CSF levels of 5-hydroxyindolacetic acid and by an apparent clinical improvement following the administration of the serotonin precursor 5-hydroxytryptophan [57, 58].
3. Altered amino acid concentrations in autopsy brains of Lesch-Nyhan patients as compared with control brains (threonine, serine, valine, isoleu-

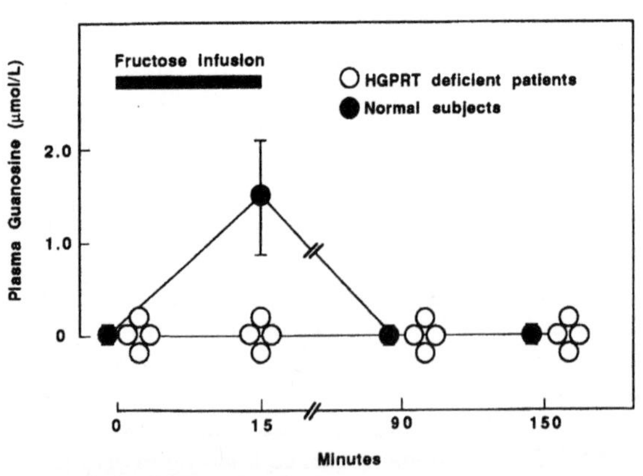



**Fig. 6.** Plasma guanosine concentrations in response to intravenous fructose infusion in normal subjects and in HGPRT deficient patients. The increase in plasma guanosine levels in four normal subjects (●) was (mean ± SEM) 1.5 ± 0.6 µmol/l. Plasma guanosine concentrations remained undetectable in four patients with HGPRT deficiency (O) following the infusion of fructose (Adapted from [13])

cine, lysine, and arginine markedly decreased; glutamine and urea higher than normal; and glutamate $\gamma$-aminobutyrate and cystathione essentially normal) [59].

4. Detectable Z nucleotide levels in erythrocytes of patients with HGPRT deficiency (45). Although lower than in HGPRT deficient patients, Z nucleotides were also present in the erythrocytes of a patient with PNP deficiency and in a patient with PRPP synthetase overactivity [45].

5. Blink reflexes were markedly different than in controls suggesting a state of hyperexcitability of the motoneurons and uncrossed interneurons and a state of hypoexcitability of the crossed interneurons at the brainstem [60].

The role of other neurotransmitter systems, such as peptidergic and purinergic neurons, have not been evaluated in HGPRT deficiency. These systems may be crucial for normal CNS function. In fact, it has been proposed that purines may act as endogenous ligands for benzodiazepine [61, 62] and adenosine [63] receptors in the brain. We believe that further collection of clinical and biochemical data in humans [64] and animal models of HGPRT deficiency is necessary before we can put together all the pieces of the puzzle that constitute the physiological and pathophysiological knowledge on neurotransmission. Observational and experimental studies may enable us in the long run to fully understand the neurological symptoms of this devastating disease.

*Acknowledgements.* We are indebted to Javier Díaz and Mrs. Mª Paz Canencia for valuable technical assistance and to Erik Lundin for help in the preparation of the manuscript. This work was supported by grants from Caja de Madrid and Fondo de Investigaciones Sanitarias de la Seguridad Social (FISS, 92/0622), Spain.

# References

1. Emmerson BT, Thompson L (1973) The spectrum of hypoxanthine-guanine phosphoribosyltransferase deficiency. Q J Med 42: 423–440
2. de Bruyn CHMM (1976) Hypoxanthine-guanine phosphoribosyl transferase deficiency. Hum Genet 31: 127–150
3. Stout JT; Caskey CT (1989) Hypoxanthine phosphoribosyltransferase deficiency: the Lesch-Nyhan syndrome and gouty arthritis. In: Scriver CR, Beaudet AL, Sly WS, Valle D (eds) The metabolic basis of inherited disease, 6th edn. McGraw-Hill, New York, pp 1007–1028
4. Lesch M, Nyhan WL (1964) A familial disorder of uric acid metabolism and central nervous system function. Am J Med 36: 561–570
5. Cristie R, Bay C, Kaufman IA, Bakay B, Borden M, Nyhan WL (1982) Lesch-Nyhan disease: clinical experience with nineteen patients. Develop Med Child Neurol 24: 293–306
6. Watts RWE, Spellacy E, Gibbs DA, Allsop J, McKeran RO, Slavin GE (1982) Clinical, post-mortem, biochemical and therapeutic observations on the Lesch-Nyhan syndrome with particular reference to the neurological manifestations. Q J Med 51: 43–78

7. Kelley WN, Greene ML, Rosenbloom FM, Henderson JF, Seegmiller JE (1969) Hypoxanthine-guanine phosphoribosyltransferase deficiency in gout. Ann Intern Med 70: 155–206

8. Seegmiller JE (1989) Contributions of Lesch-Nyhan syndrome to the understanding of purine metabolism. J Inher Metab Dis 12: 184–196

9. Nuki G, Lever J, Seegmiller JE (1974) Biochemical characteristics of 8-azaguanine resistant human lymphoblast mutants selected in vitro. Adv Exp Med Biol 41A: 255–267

10. Seegmiller JE, Rosenbloom FM, Kelley WN (1967) Enzyme defect associated with a sex-linked human neurological disorder and excessive purine synthesis. Science 155: 1682–1684

11. Sorensen LB (1970) Mechanism of excessive purine biosynthesis in hypoxanthine-guanine phosphoribosyltransferase deficiency. J Clin Invest 49: 968–978

12. Edwards NL, Recker D, Fox IH (1979) Overproduction of uric acid in hypoxanthine-guanine phosphoribosyltransferase deficiency. Contribution by impaired purine salvage. J Clin Invest 63: 922–930

13. Puig JG. Jiménez ML, Mateos FA, Fox IH (1989) Adenine nucleotide turnover in hypoxanthine-guanine phosphoribosyl-transferase deficiency: evidence for an increased contribution of purine biosynthesis de novo. Metabolism 38: 410–418

14. Puig JG, Mateos FA, Jiménez ML, Ramos TH (1988) Renal excretion of hypoxanthine and xanthine in primary gout. Am J Med 85: 533–537

15. Rosenbloom FM, Henderson JF; Caldwell IC, Kelley WN, Seegmiller JE (1968) Biochemical basis of accelerated purine biosynthesis de novo in human fibroblasts lacking hypoxanthine-guanine phosphoribosyltransferase. J Biol Chem 243: 1166–1173

16. Itakura M, Sabina RL, Heald PW, Holmes EW (1981) Basis for the control of purine biosynthesis by purine ribosnucleotides. J Clin Invest 67: 994–10002

17. Zoref-Shani E, Sperling O (1980) Dependence of the metabolic fate of IMP on the rate of total IMP synthesis. Studies in cultured fibroblasts from normal subjects and from purine-overproducing mutant patients. Biochim Biophys Acta 607: 503–511

18. Vincent MF, van der Berghe G, Hers HG (1986) Effect of fructose on the concentration of phosphoribosyltransferase in isolated hepatocytes. Adv Exp Med Biol 195B: 615–621

19. Nuki G, Astria K, Brenton DP, Cruikshank MK, Lever J, Seegmiller JE (1977) Purine and pyrimidine concentration in cells with decreased hypoxanthine-guanine phosphoribosyltransferase activity. Adv Exp Med Biol 76A: 326–339

20. Brenton DP, Astria K, Cruikshank MK, Seegmiller JE (1977) Measurement of free nucleotides in cultured human lymphoid cells using high pressure liquid chromatography. Biochem Med 17: 271–274

21. Lommen EJP, Vogels GD, van der Zee SPM, Trijbels JMF, Schretlen EDAM (1971) Concentrations of purine nucleotides in erythrocytes of patients with the Lesch-Nyhan syndrome before and during oral administration of adenine. Acta Pediat Scand 60: 642–646

22. Rivard GE, Izadi P, Lazerson J, McLaren JD, Parker C, Fish CH (1975) Functional and metabolic studies of platelets from patients with Lesch-Nyhan syndrome. Br J Hematol 31: 245–253

23. Wood AW, Becker MA, Seegmiller JE (1973) Purine nucleotide synthesis in lymphoblasts cultured from normal subjects and a patient with Lesch-Nyhan syndrome. Biochem Genet 9:261–274

24. Holmes EW, Wyngaarden JB, Kelley WN (1973) Human glutamine phosphoribosyl-pyrophosphate amidotransferase: two molecular forms interconvertible by purine ribonucleotides and phosphoribosylpyrophosphate. J Biol Chem 248: 6035–6040
25. Green ML, Boyle JA, Seegmiller JE (1970) Substrate stabilization: genetically controlled reciprocal relationship of two human enzymes. Science 167: 887–889
26. Rubin CS, Balis ME, Pionelli S, Berman PH, Dancis J (1969) Elevated AMP pyro-phosphorilase activity in congenital IMP pyrophosphorilase deficiency (Lesch-Nyhan syndrome). J Lab Clin Med 74: 732–741
27. Gordon RB, Thompson L, Emmerson BT (1974) Erythrocyte phosphoribosylpyro-phosphate concentrations in heterozygotes for hypoxanthine-guanine phosphoribosyl-transferase deficiency. Metabolism 23: 921–927
28. Emmerson BT, Gordon RB (1986) HGPRT deficiency with normal erythrocyte PRPP and APRT activity. Adv Exp Med Biol 195A: 163–165
29. Watts RWE, Watts JEM, Seegmiller JE (1965) Xanthine oxidase activity in human tissues and its inhibition by allopurinol. J Lab Clin Med 66: 688–697
30. Jarasch E-D, Grund C, Bruder G, Heid HW, Keenan TW, Franke WW (1981) Localization of xanthine oxidase in mammary-gland epithelium and capillary endothelium. Cell 25: 67–82
31. Wilson JM, Young AB, Kelley WN (1983) Hypoxanthine-guanine phosphoribosyltransfer-ase deficiency. The molecular basis of the clinical syndromes. N Eng J Med 309: 900–910
32. Sass JK, Itabashi HH, Dexter RA (1965) Juvenile gout with brain involvement. Arch Neurol 13: 639–655
33. Crussi FG, Robertson DM, Hiscox JL (1969) The paradoxical condition of the Lesch-Nyhan syndrome: report of two cases. Am J Dis Child 118: 501–506
34. Mizuno T, Endoh H, Konishi Y, Miyachi Y, Akaoka I (1976) An autopsy case of the Lesch-Nyhan syndrome: normal HGPRT activity in liver and xanthine calculi in various tissues. Neuropaediatrie 76: 351–355
35. Allsop J, Watts RWE (1980) Activities of amido phosphoribosyltransferase and purine phosphoribosyltransferase in developing rat brain. Adv Exp Med Biol 122A: 361–366
36. Rosenbloom FM, Kelley WN, Miller J, Henderson JF, Seegmiller JE (1967) Inherited disorder of purine metabolism: correlation between central nervous system dysfunc-tion and biochemical defects. JAMA 202: 175–177
37. Lloyd KG, Hornykiewicz O, Davidson L, Shannak K, Farley I, Goldstein M, Shibuya M, Kelley WN, Fox IH (1981) Biochemical evidence of dysfunction of brain neuro-transmitters in the Lesch-Nyhan syndrome. N Eng J Med 305: 1106–1011
38. Rijksen G, Staal GEJ, van der Vlist MJM (1981) Partial hypoxanthine-guanine phos-phoribosyltransferase deficiency with full expression of the Lesch-Nyhan syndrome. Hum Genet 57: 39–47
39. Page T, Bakay B, Nisinen E, Nyhan WL (1981) Hypoxanthine-guanine phosphoribosyl-transferase variants: correlation of clinical phenotype with enzyme activity. J Inherit Metab Dis 4: 203–206
40. Al-Khalidi USA, Chaglassian TH (1965) The species distribution of xanthine oxidase. Biochem J 97: 316–320
41. Sweetman L (1968) Urinary and cerebrospinal fluid oxypurine levels and allopurinol metabolism in the Lesch-Nyhan syndrome. Fed Proc 27: 1055–1059
42. Hoefnagel D 81967) Clinical features of the Lesch-Nyhan syndrome: pathology and pathologic physiology. Fred Proc 27: 1042–1046
43. Jiménez ML, Puig JG, Antón FM, Hernández TR, Castroviejo IP, Vázquez JO (1989) Transporte de purinas a través de la barrera hematoencefálica en la deficiencia de hipo-xantina fosforribosiltransferasa. Med Clin (Barc) 92: 167–170

44. Howard WJ, Kerson LA, Appel SH (1970) Synthesis de novo of purines in slices of rat brain and liver. J Neurochem 17: 121–128
45. Sidi Y, Mitchell BS (1985) Z-nucleotide accumulation in erythrocytes from Lesch-Nyhan patients. J Clin Invest 76: 2416–2419
46. Greene I, Urdin IB, Synder SH (1979) Dopamine receptor binding regulated by guanine nucleotides. Mol Pharmacol 16: 69–76
47. Goldstein M, Anderson LT, Reuben R, Dancis J (1985) Self-mutilation in Lesch-Nyhan disease is caused by dopaminergic denervation [letter]. Lancet 1: 338–339
48. Ungerstedt U (1971) Postsynaptic supersensitivity after 6-hydroxydopamine induced degeneration of the nigrostriatal dopamine system. Acta Physiol Scand Suppl 361: 69–91
49. Goldstein M, Kuga S (1984) Dopamine agonist induced compulsive biting behavior in monkeys. Animal model for Lesch-Nyhan syndrome [abstract]. Soc Neurosci 239.
50. Goldstein M, Kuga S (1984) Possible involvement of central dopamine receptors in compulsive self-mutilative behavior [abstract]. Am Neurol Assoc P39
51. Goldstein M, Anderson LT, Reuben R, Dancis J (1985) Self-mutilation in Lesch-Nyhan disease is caused by dopaminergic denervation [letter]. Lancet 1: 338
52. Simmonds HA, Fairbanks LD, Morris GS, Webster DR, Harley EH (1988) Altered erythrocyte nucleotide patterns are characteristic of inherited disorders of purine or pyrimidine metabolism. Clin Chim Acta 171: 197–210
53. Mateos FA, Puig JG, Jiménez ML, Fox IH (1987) Hereditary xanthinuria. Evidence for enhanced hypoxanthine salvage. J Clin Invest 79: 847–52
54. Simmonds HA, Webster DR, Wilson J, Potter CF, Fairbanks LD (1984) Evidence of a new syndrome involving hereditary uric acid overproduction, neurological complications and deafness. Adv Exp Med Biol 165A: 97–102
55. Rockson S, Stone R, van der Weyden M, Kelley WN (1974) Lesch-Nyhan syndrome: evidence for abnormal adrenergic function. Science 186: 934–935
56. Lake CR, Ziegler MG (1977) Lesch-Nyhan syndrome: low dopamine-$\beta$-hydroxylase activity and diminished sympathetic response to stress and posture. Science 196: 905–906
57. Castells S, Chakrabarti C, Winsberg BG, Hurwic M, Perel JM, Nyhan WL (1979) Effects of L-5-hydroxytryptophan on monoamine and amino acids turnover in the Lesch-Nyhan syndrome. J Autism Dev Disord 9: 95–103
58. Anders TF, Cann HM, Ciaranello RD, Barchas JD, Berger PA (1978) Further observations on the use of 5-hydroxytryptophan in a child with Lesch-Nyhan syndrome. Neuropediatrie 9: 157–166
59. Rassin DK, Lloyd KG, Kelley WN, Fox IH (1982) Decreased amino acids in various brain areas of patients with Lesch-Nyhan syndrome. 13: 130–134
60. Hatanaka T, Higashino H, Woo M, Yasuhara A, Sugimoto T, Kobayashi Y (1990) Lesch-Nyhan syndrome with delayed onset of self-mutilation: hyperactivity of interneurons at the brainstem and blink reflex. Acta Neurol Scand 81: 184–187
61. Skolnick P, Paul SM, Marangos PJ (1980) purines as endogenous ligands of the benzodiazepine receptor. Fed Proc 39: 3050–3055
62. Phillis JW (1979) Diazepam potentiation of purinergic depression of central neurons. Can J Physiol Pharmacol 57: 432–435
63. Daly JW; Burns RF, Snyder SH (1981) Adenosine receptors in the central nervous system: relationship to the central actions of methylxanthines. Life Sci 28: 1083–2097
64. Mizuno T (1986) Long-term follow-up of ten patients with Lesch-Nyhan syndrome. Neuropediatrics 17: 158–161

# 4 Prenatal Diagnosis of Lesch-Nyhan Syndrome

F. A. Mateos and J. G. Puig

Lesch-Nyhan syndrome (LNS) is an X-linked recessive disorder that is transmitted by asymptomatic carrier females [1, 2]. Except for one case of a female with the LNS [3], in every family reported, transmission of the disease has been through the female to the affected male. The devastating neurological symptoms of this incurable disease makes its prevention of paramount importance. Prevention of LNS can be sought by means of detection of female carriers for hypoxanthine guanine phosphoribosyltransferase (HGPRT) deficiency and prenatal diagnosis. In families with one child or more affected by LNS, potential carriers must be screened for a deficient activity of HGPRT. The theoretical chance that a male fetus of a pregnant female carrier of LNS suffers the enzymatic defect is 50% and female carriers may ask for information about the risk of having offspring with this genetic disease. Antenatal diagnosis must be offered early because if an abortion is chosen it should take place at a stage of pregnancy when maternal – fetal bonding is less.

The first prenatal diagnosis of LNS during the first trimester of pregnancy was made in 1968 [3]. This study used a radioautographic procedure showing the incorporation of [$^3$H] hypoxanthine into the nucleic acid of cells cultured from the amniotic fluid. Since then, some authors have published the results of prenatal diagnosis of LNS by measuring HGPRT activity in amniotic cells and in chorionic villus samples [4–8]. More recently, the prenatal diagnosis of LNS has been documented by DNA analysis of extracts from cultured amniocytes or chorionic villus samples [9, 10]. We recently reported the antenatal diagnosis of a fetus with LNS by analysis of HGPRT activity in cord blood [11].

Since chorionic villus sampling, culture of amniotic cells, and cordocentesis are alternative procedures for prenatal diagnosis of LNS, the diagnostic accuracy, risk for the fetus, and possibility of earlier diagnosis need to be considered (Table 1).

Amniocentesis may be performed between the 15th and 17th gestational weeks [12, 13]. With this procedure the results can be provided within the 18th to 20th gestational week, and the risk of fetal loss is small (0.5%–1.5%). However, in some cases, the amniotic cell cultures may be contaminated with maternal cells causing a fallacious mosaicism [14]. The chorionic villus biopsy permits an earlier diagnosis (8th–12th gestational week) [12] but possibility of fetal loss is

**Table 1.** Methods for the prenatal diagnosis of Lesch-Nyhan syndrome

| Method | Stage of gestation | Fetal loss (%) |
| --- | --- | --- |
| Chorionic villus sampling | 8–12 | 2.3–3.4 |
| Amniocentesis | 15–17 | 0.5–1.5 |
| Cordocentesis | 18–20 | 1.9–2.0 |

higher (2.3%–3.4%) and the sample can be mixed with maternal cells resulting in an inaccurate diagnosis [14–16]. Umbilical cord puncturing is a relatively new diagnostic procedure that can be performed in the 18th–20th gestational week. The chance of fetal death with this technique is close to 2% [17]. This procedure is accurate because of the possibility of identifying the origin of the sample in situ by acid elution of fetal hemoglobin. However, puncturing the umbilical cord in the early stage of pregnancy can be extremely difficult.

We have made one prenatal diagnosis of LNS by cordocentesis [11]. In this study we documented high oxypurine levels in amniotic fluid. Our patient, a pregnant woman age 26, was known to be a carrier of the LNS and requested antenatal diagnosis in her 20th gestational week. The sex of the fetus (male) was determined by a chromosome study during the 18th gestational week. She had a cousin and two second cousins that suffered the Lesch-Nyhan syndrome. Two brothers died as a consequence of an undiagnosed illness with the clinical characteristics of the Lesch-Nyhan syndrome (Fig. 1). In the 20th gestational week, an umbilical cord puncture was performed transabdominalli and 2 ml of blood and 2 ml of amniotic fluid were obtained. Using a drop of blood, we performed the Nierhaus and Betke reaction. The sample belonged to the mother. We simultaneously measured uric acic, hypoxanthine, and xanthine concentrations in the amniotic fluid, and we found that these purine compounds were increased about 2, 2.8, and 3 times, respectively, when compared to the observed values in 14 amniotic fluids from normal pregnancies (Table 2). Four days later, a second cordocentesis was performed and the blood sample belonged to the fetus. HGPRT activity was undetectable and adenine phosphoribosyltransferase (APRT) was increased. HGPRT and APRT activities in the pregnant woman were similar to those obtained in 50 normal controls. The diagnosis of LNS was established. At the request of the mother, the pregnancy was terminated and 2 ml of fetal blood by cardiac puncture were taken. HGPRT and APRT enzymatic activities in fetal erythrocytes obtained post-abortion were similar to those obtained in cord blood (Fig. 2).

Umbilical cord puncturing is an accurate method for the prenatal diagnosis of diseases with some expression in fetal blood [12, 17]. However the technique is limited by the difficulty in puncturing the umbilical cord. In our patient we attempted cordocentesis in the 20th gestational week. Although we could not get fetal blood, we obtained amniotic fluid and postulated that the analysis of oxypurines could provide information about the HGPRT enzyme activity of the

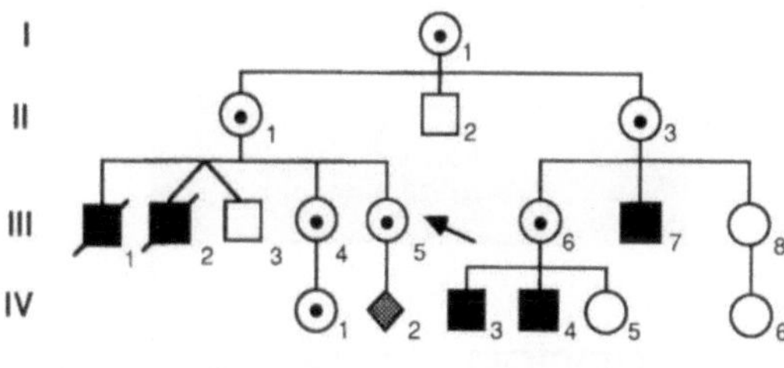

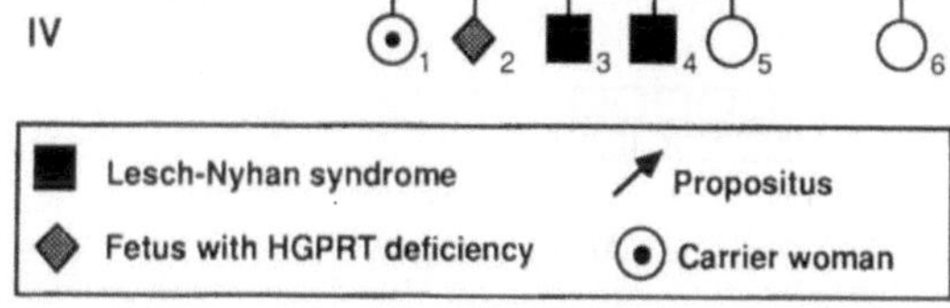


**Fig. 1.** Predigree of a family with the Lesch-Nyhan syndrome

**Table 2.** Oxypurine measurements in 14 normal pregnant women and in the amniotic fluid, plasma, and 24 h urine of a female carrier of the Lesch-Nyhan syndrome

|  | Normal amniotic fluid | Problem amniotic fluid | Female carrier of LNS Plasma | 24 h urine |
|---|---|---|---|---|
| Gestational week | 17 ± 3[a] | 20 | | |
| Uric acid (µmol/l) | 1547 ± 357 | 2975 | 321 | |
| Hypoxanthine (µmol/l) | 0.6 ± 0.5 | 4.5 | 3.0 | |
| Xanthine (µmol/l) | 1.0 ± 0.4 | 5.2 | 1.2 | |
| Uric acid (µmol/g creatinine) | 20.8 ± 4.9 | 39.2 | | 3309 |
| Hypoxanthine (µmol/g creatinine) | 74 ± 49 | 600 | | 59 |
| Xanthine (µmol/g creatinine) | 135 ± 54 | 680 | | 45 |

[a] Values are the means ± SD.

fetus. In the first half of a normal pregnancy, the amniotic fluid composition is similar to the fetal extracellular fluid, and its study has been used for the diagnosis of inherited diseases [18]. An increase of purine nucleotide degradation is readily recognized by an increase of nucleosides and purine bases in biological fluids [19]. Few studies have examined the normal concentration of amniotic fluid oxypurines in the first gestational trimester [20]. However, some authors have demonstrated than in fetal hypoxia there is an increase of hypoxanthine, xanthine, and uric acid levels in cord blood and in amniotic fluid [21, 22]. In HGPRT deficiency, lack of hypoxanthine reutilization and acceleration of de novo purine synthesis explains the increase of oxypurines in body fluids [2]. The amniotic fluid examined by us showed a uric acid concentration above the mean plus two standard deviations above the observed values in 14 control amniotic fluids from normal pregnancies with a similar gestational age. In addition, hypo-

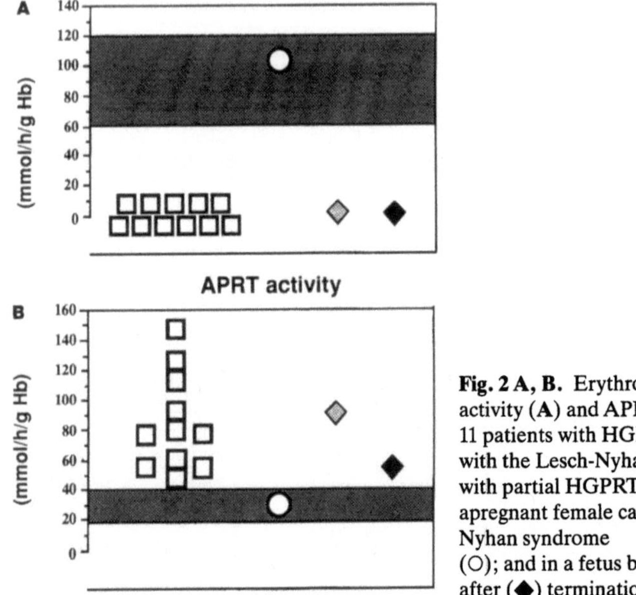



**Fig. 2 A, B.** Erythrocyte HGPRT activity (**A**) and APRT activity (**B**) in 11 patients with HGPRT deficiency, 8 with the Lesch-Nyhan syndrome and 3 with partial HGPRT activity (□); in apregnant female carrier of the Lesch-Nyhan syndrome (○); and in a fetus before (◇) and after (◆) termination of pregnancy

xanthine and xanthine amniotic fluid concentrations were 3.5 and 2.8 times the upper normal limit for control amniotic fluids. These findings in a fetus which was later confirmed to be deficient for HGPRT activity suggest that amniotic fluid analysis in the early stages of pregnancy can be useful for the prenatal diagnosis of the LNS. The possibility that the increased amniotic fluid oxypurines could be the result of maternal increased oxypurine levels transferred to the fetus (Sperling, personal communication) is ruled out by the fact that simultaneously maternal plasma hypoxanthine and xanthine were 3.0 and 1.2 u$M$, respectively, and 24 h urinary hypoxanthine and xanthine concentrations were 59 and 45 µmol/g creatinine, respectively (Table 2). Considering the possibility of contamination of amniotic cell cultures and chorionic villus biopsy by maternal cells [23, 24], analysis of oxypurine concentrations in amniotic fluid represent a new technique for the early diagnosis of LNS. Due to the small volume of the sample needed and the short time required for oxypurine analysis (roughly 2 h), this method could be regarded as a valuable alternative in the early diagnosis of inherited purine metabolic diseases.

*Acknowledgements*   We are indebted to Javier Díaz and Mrs. M$^a$ Paz Canencia for valuable technical assistance and to Erik Lundin for help in the preparation of the manuscript. This work was supported by grants from Caja de Madrid and Fondo de Investigaciones Sanitarias de la Seguridad Social (FISS, 92/0622), Spain.

# References

1. Lesch M, Nyhan WL (1964) A familial disorder of uric acid metabolism and central nervous system function. Am J Med 36: 561–570
2. Stout JT, Caskey CT (1989) Hypoxanthine phosphoribosyltransferase deficiency: the Lesch-nyhan syndrome and gouty arthritis. In: Scriver CR, Beaudet AL., Sly WS, Valle D (eds) The metabolic basis of inherited disease, 6th edn. McGraw-Hill, New York, pp 1007–1028
3. Fujimoto WY, Seegmiller JE, Uhlendorf BW, et al (1968) Bichemical diagnosed of an X-linked disease in utero. Lancet ii; 511–512
4. Boyle JA, Raivio KO, Astrin KH, et al (1970) Lesch-Nyhan syndrome: preventative control by prenatal diagnosis. Science 169: 688–689
5. Crawhall JC, Henderson JF, Kelley WN (1972) Diagnosis and treatment of the Lesch-Nyhan syndrome. Pediatr Res 6: 504–513
6. De Bruyn CHMM. (1976) Hypoxanthine-guanine phosphoribosyltransferase deficiency in man. Hum Genet 31: 127–150
7. Gibbs DA, McFadyen IR, Crawfurd MD'A, et al (1984) First trimester prenatal diagnosis of Lesch-Nyhan syndrome. Lancet ii: 1180–1183
8. Stout JT, Jackson LC, Caskey CT (1985) First trimester diagnosis of Lesch-Nyhan syndrome: application to other disorders of purine metabolism. Prenat Diagn 5: 183–189
9. Pai GS, Sprenke JA, Do TT, Mateni CE, Migeon RO (1980) Localization of loci for hypoxanthine phosphoribosyltransferase and glucose-6-phosphate dehydrogenase and biochemical evidence of non-random X-chromosome expression from human X-autosomal translocation. Proc Natl Acad Sci USA 77: 2810–2813
10. Singh S, Willer I (1991) Biochemical and molecular genetic investigation of HPRT deficiency mutations in a Turkish and three German families: heterozygote, prenatal and postnatal diagnosis with cell culture, DNA blot and PCR technique (Abstract). Int J Purine Pyrimidine Res 2(Suppl 1): 86
11. Mateos FA, Puig TH, Jiménez ML, Romera NM, Gonzáles A (1991) Prenatal diagnosis of Lesch-Nyhan syndrome by purine analysis of amniotic fluid and cordocentesis. Adv Exp Med Biol 309 B: 47–50
12. Valle D (1991) Treatment and prevention of genetic disorders. In: Wilson JD, Braunwald E, Isselbacher KJ, Petersdordorf RG, Martin JB, Fauci AS, Root RK (eds) Principles of internal medicine, 12th edn. McGraw-Hill, New York, pp 55–60
13. Gibs DA, Headhouse-Benson CM, Watts RWE (1986) Family studies of the Lesch-Nyhan syndrome: the use of restriction fragment length polymorphism (RFLP) closely linked to the disease gene for carrier state and prenatal diagnosis. J Inher Dis 9: 45–58
14. Canadian Collaborative CVS-Amniocentesis Clinical Trial Group (1989) Multicentre randomized clinical trial of chorion villus sampling and amniocentesis. Lancet i: 1–6
15. Rhoads GG, Jackson LG, Schlesselman SE, et al (1989) The safety and efficacy of chorionic villus sampling for early prenatal diagnosis of cytogenetic abnormalities. N Eng J Med 320: 609–617
16. Gruber A, Zeitune M, Fejgin M (1989) Failure to diagnose Lesch-Nyhan syndrome by first trimester chorionic villus sampling. Prenat Diag 9: 452–453
17. Daffos F, Capella-Pavlovsky M, Forestier FA (1983) A new procedure for fetal blood sampling in utero: preliminary results of 53 cases. Am J Obstet Gynecol 146: 985–987
18. Seeds AE (1980) Current concepts of amniotic fluid dynamics. Am J Obstet Gynecol 138: 575–586

19. Fox IH (1985) Adenosine triphosphate degradation in specific disease. J Lab Clin Med 106: 101–110
20. Harkness RA (1989) Hypoxanthine, xanthine and uridine in body fluids, indicators of depletion. J Chromatogr Biom Appl 429: 255–278
21. Saugstad OD (1988) Hypoxanthine as an indicator of hypoxia. Its role in health and disease through free radical production. Pediatr Res 23: 143–150
22. Mateos FA, JG. Puig JG, Ramos TH, et al (1989) Erythrocyte ATP (iATP) as an indicator of neonatal hypoxia. Adv Exp Med Biol 253A: 345–352
23. Aula P, Matilla K, Pironinen O, et al (1989) First trimester prenatal diagnosis of aspartylglucosaminuria. Prenat Diag 9: 617–620
24. Benn PA, Hsu LYF (1983) Maternal cell contamination of amniotic fluid cell cultures. Results of a US nationwide survey. Am J Med Genet 15: 297–305

# 5  The Genetic Basis of HGPRT Deficiency

B. L. DAVIDSON and B. J. ROESSLER

Complete deficiency of the purine salvage enzyme hypoxanthine guanine phosphoribosyltransferase (HGPRT) results in the Lesch-Nyhan syndrome, a disease characterized by hyperuricemia, gout, and a bizarre tendency to self-mutilation, choreoathetosis, and other neurological dysfunction [1]. Partial deficiency of HGPRT is characterized by gout, hyperuricemia, and hyperuricaciduria and is known as the Kelley-Seegmiller syndrome [2]. The gene encoding HGPRT is located on the long arm of the X-chromosome and most often transmitted in a classic X-linked manner with an incidence of 1:100 000 live births. However, spontaneous cases do occur and at least one case of an affected female has been reported [3].

The genetic analysis of HGPRT deficiency states represents a triumph of biomedical science. The discovery of HGPRT deficiency as an inherited form of gout has provided an important model for the study of inborn errors of metabolism. Major accomplishments have been made from the time of the initial clinical description of the Lesch-Nyhan syndrome to the elucidation of its biochemical and enzymatic basis and to the definition of its molecular basis at the level of DNA.

The genetic analysis of HGPRT deficiency states has proceeded rapidly in the past few years and to date greater than 70 mutations associated with HGPRT deficiency states have been defined on a molecular level [4–9]. The molecular basis of HGPRT deficiency can be studied using multiple methods. At the protein level abnormalities in HGPRT can be evaluated by employing methods that include electrophoretic mobility assays, kinetic studies, enzyme activity assays, and study of immunoreactive material. Genetically, the use of conventional cDNA cloning, polymerase chain reaction (PCR) amplification, and nucleotide sequencing have allowed a detailed analysis of the mutations responsible for HGPRT deficiency states. Implications as to the impact of genetic mutations on the structure-function aspects of the HGPRT molecule can be predicted through the use of computer analysis. Unfortunately, these methods are imprecise and therefore a final goal, yet to be realized, is the crystallographic analysis of normal and mutant HGPRT proteins.

Data obtained from HGPRT activity assays and immunoblots for HGPRT cross-reactive material (CRM) in lymphoblasts derived from a series of patients with partial and complete HGPRT deficiency have shown the following: Most Lesch-Nyhan patients lack detectable HGPRT activity ($< 0.7\%$ of normals) and show significantly reduced levels of CRM. In contrast, partially deficient subjects generally show a less pronounced reduction in the levels of HGPRT activity with levels of CRM ranging from approximately $1\%$ to $50\%$ of normals.

The etiology of the diminished levels of HGPRT protein produced in these individuals has previously been evaluated by Northern analysis of total RNA isolated from lymphoblasts derived from affected patients. These blots were probed for human HGPRT mRNA and co-probed for adenine phosphoribosyl-transferase (APRT) mRNA as an internal control. The HGPRT mRNA levels are grossly normal in most variants. Several mutants including WE, GM2292, and BD have no detectable levels of HGPRT mRNA. A single cell line, GM6804, reveals an abnormally large HGPRT mRNA molecule that has been shown to be the result of an internal duplication of exons 2 and 3 [10].

In summary, most patients with partial HGPRT deficiency exhibit normal levels of HGPRT mRNA and near normal levels of CRM, while those exhibiting complete deficiency of HGPRT exhibit markedly reduced amounts of immuno-reactive HGPRT protein; a few lack HGPRT mRNA entirely. This information suggests that the majority of genetic mutations responsible for HGPRT defi-ciency states represent point mutations within the coding regions and are not the result of gross rearrangements occurring in the HGPRT gene. Support for this hypothesis was obtained by Southern analysis of the HGPRT gene in affected individuals. The results of Southern blot analysis using two restriction endo-nucleases (*Bam*HI and *Taq*I) show no gross changes in the structure of the HGPRT genome in a large number of affected patients.

Initially, mutational analysis relied on amino acid sequencing of purified HGPRT and/or conventional techniques for the cloning of HGPRT cDNA. Se-veral HGPRT mutants have been identified through the use of these techniques. They include $HGPRT_{Kingston}$, $HGPRT_{London}$, $HGPRT_{Munich}$, $HGPRT_{Toronto}$ identified by amino acid sequencing, and $HGPRT_{London}$, $HGPRT_{Ann\ Arbor}$, $HGPRT_{Flint}$, $HGPRT_{Midland}$, $HGPRT_{Yale}$ and $HGPRT_{Ashville}$ identified by cDNA cloning [4, 5, 8]. In most cases, however, the amino acid sequence analysis gives only indirect evidence for a specific genetic mutation and must be confirmed by direct genetic analysis. Additionally, these methods are inefficient and labo-rious, especially for use in identifying larger numbers of mutations. Fortunately, in the past few years advances in the techniques of molecular biology have allowed investigators to rapidly and directly determine the genetic mutations responsible for alterations in HGPRT enzyme function.

One of the most important developments that has facilitated the genetic ana-lysis of HGPRT deficiency states has involved the use of PCR [11]. This method allows the production of large amounts of HGPRT cDNA that in turn can be di-rectly sequenced or easily cloned into sequencing vectors. The use of direct se-quencing of PCR amplified transcripts has been a major focus of our labora-tory during the past several years for the analysis of HGPRT mutants. The technique is outlined as follows: Briefly, whole cellular RNA is subjected to oligo d(T) priming and reverse transcription. This is followed by PCR amplification for 30 cycles using HGPRT-specific sense and antisense flanking primers and thermostable DNA polymerase [11]. The flanking oligonucleotide primers contain sequences from the 5'and 3'untranslated regions of the HGPRT mRNA and allow the generation of a partial length HGPRT cDNA containing the start codon, stop codon, and the entire protein coding region. The partial length HGPRT cDNA amplified by PCR can be examined by agarose gel electrophoresis and appears as a single band of 760 base pairs (bp).

The HGPRT cDNAs are then directly sequenced using a modification of the dideoxy chain termination method and a series of overlapping HGPRT specific oligonucleotide primers labeled with $^{32}$P [7, 8]. As a technical point, the oligonucleotide sequencing primers must be nested within the flanking oligonucleotide primers used for PCR amplification. We have used a series of eight unique HGPRT-specific oligonucleotide primers (between 17 and 20 bp in length) designed to provide large regions of overlapping sequence information in both orientations of the HGPRT cDNA.

When performing mutational analysis using this technique it is necessary to confirm the identity of the mutation with alternative methods. For example, reverse transcriptase and thermostable DNA polymerases are subject to base substitution errors at a rate of approximately $1 \cdot 10^{-4}$ and $1 \cdot 10^{-5}$ respectively [11, 12]. Thus additional methods are used to confirm that a base pair alteration identified by direct sequencing of PCR amplified transcripts represents the true HGPRT mutation and not an artifact of the molecular methods employed. Techniques commonly used to confirm mutations include amino acid sequencing, Southern analysis, RNase protection mapping, allele-specific hybridization, and direct genomic sequencing with or without multiplex sequencing [4–9].

By using these techniques our laboratory has identified a large number of point mutations that result in the production of abnormal HGPRT proteins [7–9]. The results of these studies are shown in Fig. 1. As can be seen from this illustration the point mutations tend to cluster within several regions, specifically exons 3, 5 and 8. This suggest that these regions of the HGPRT gene represent mutational hot spots. The importance of this observation, in terms of defining gene stability and determining the mechanisms underlying mutational events, has yet to be systematically evaluated. The continued use of normal and mutant HGPRT genes will provide an important model system for the study of these questions.

The definition of these point mutations has had an important impact on the study of the structure-function aspects of the HGPRT enzyme. Computer modeling can be used to produce a conceptualized three-dimensional representation of the HGPRT molecule showing the location of defined point mutations in relationship to the putative hypoxanthine and 5-phosphoribosyl-1-pyrophosphate (PRPP) binding regions of the enzyme. Additionally, structure algorithms and computer modeling can make predictions regarding the effect of specific point mutations on protein secondary structure [13, 14]. Examples of point mutations predicting a variety of structural changes include HGPRT$_{Banbury}$

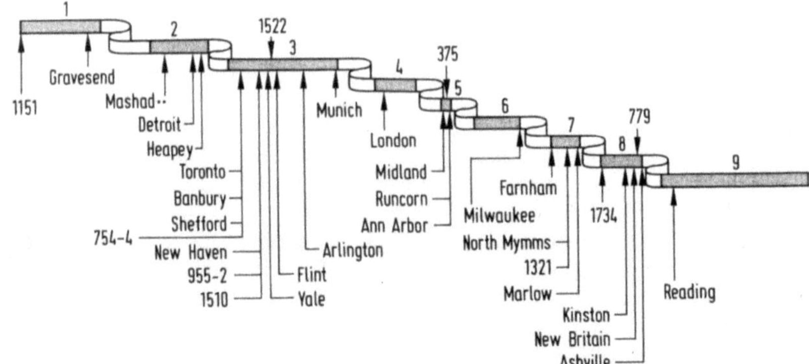



**Fig. 1.** The point mutations identified in the human HGPRT gene. The linear representation of the HGPRT gene shows in *red* and introns in *white*. The locations of point mutations are shown by *yellow arrows*, and if known, identified by the geographic location of the affected kindred

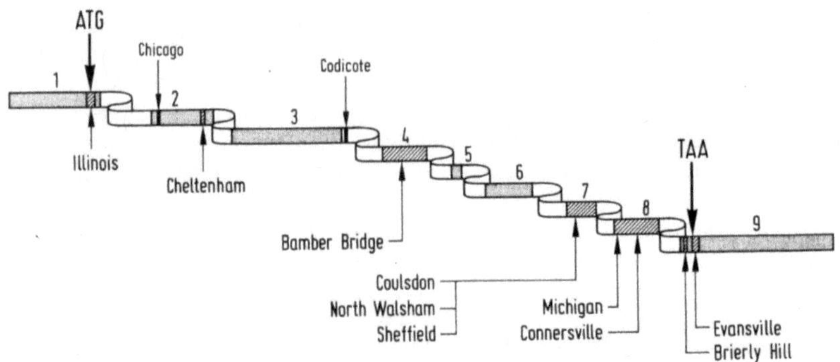



**Fig. 2.** The deletions and insertions identified in the human HGPRT gene. The linear representation of the HGPRT gene shows exons in *red* and introns in *white*. the locations of deletions are denoted by *yellow bars* and *arrows*. The location of insertions are denoted by *blue bars* and *narrows*. If known, each mutation is identified by the geographic location of the affected kindred

($\alpha$ helix to random coil). HGPRT$_{Munich}$ (random coil to $\beta$ sheet), HGPRT$_{Midland}$ (random coil to $\beta$ turn), HGPRT$_{Ann Arbor}$ ($\beta$ sheet to $\alpha$ helix), and HGPRT$_{Arlington}$ ($\alpha$ helix to $\beta$ sheet). More sophisticated structure determinations await the purification, crystallization, and X-ray analysis of recombinant normal and mutant human HGPRT proteins.

The techniques of direct sequencing of PCR amplified HGPRT cDNA and direct sequencing of regions of the HGPRT gene have also allowed the identification of several mutations characterized by deletions and insertions (Fig. 2). Undoubtedly many of these mutations result in the abnormal processing of HGPRT hnRNA into HGPRT mRNA. Others may accelerate HGPRT mRNA degradation or produce mRNA that cannot be translated into normal HGPRT protein. In $HGPRT_{Chicago}$ and $HGPRT_{Codicote}$, insertions create stop codons that result in premature translational termination of the HGPRT mRNA.

The molecular basis of HGPRT duplications and deletions has begun to be studied in a systematic manner. Studies by Monnat et al. suggest that some HGPRT deletions are generated by illegitimate recombinations [15]. The HGPRT genome contains large areas of *Alu* repeats and other repetitive motifs. Surprisingly, a low frequency of deletions appears to have arisen from homologous recombinational events between the repeated elements.

In order to better understand the impact of these deletions on the phenotypic expression of HGPRT activity, our laboratory has begun to analyze the genetic basis of HGPRT deficiency in two HGPRT mutants that involve alteration of the ATG initiation codon; $HGPRT_{Illinois}$, a deletion in exon 1 and $HGPRT_{1151}$, a point mutation in exon 1. Clinically, $HGPRT_{Illinois}$ is manifest as the Kelley-Seegmiller syndrome while $HGPRT_{1151}$ is manifest as the Lesch-Nyhan syndrome [6, 7]. In an attempt to understand how similar genetic alterations can result in disparate clinical disease, our laboratory has further defined the genetic basis of the functional abnormality expressed in these mutants.

A comparison of the sequence of normal HGPRT cDNA, $HGPRT_{Illinois}$ cDNA, and $HGPRT_{1151}$ cDNA is as follows: $HGPRT_{Illinois}$ is characterized by a 13 bp deletion from nucleotide $-12$ to $+1$ that results in the alteration of the ATG initiation codon to GTG. The resulting sequence is 5'... CCCGCGCGCC *GTG* ...3'. $HGPRT_{1151}$ is a G to A point mutation at nucleotide $+3$ that results in the alterations of the *ATG* initiation codon to *ATA*. In both cases, the next inframe ATG codon occurs 126 bp downstream. If $HGPRT_{Illinois}$ and $HGPRT_{1151}$ were produced from translational initiation at this second methionine codon, the expected HGPRT proteins would be shortened by 42 amino acids and would likely appear as a distinctly smaller proteins on HGPRT activity gel assay.

Cell lysates from normal and $HGPRT_{Illinois}$ derived lymphoblasts have been studied using HGPRT activity gel assays. These experiments show that normal HGPRT and $HGPRT_{Illinois}$ are identical in size. In the case of $HGPRT_{Illinois}$, the amount of HGPRT activity present in cell lysates is significantly reduced and was calculated to be less than 1% of normal. No detectable HGPRT activity is present in activity gels of $HGPRT_{1151}$ cell lysate. Immunoblots of cell lysates from $HGPRT_{Illinois}$ reveal immunoreactive HGPRT of the correct size when samples were overloaded 100 times with $HGPRT_{Illinois}$ cell lysate. The amount of immunoreactive HGPRT produced by $HGPRT_{Illinois}$ cell was approximately 1% of normal. No immunoreactive material could be detected in cell lysates of $HGPRT_{1151}$.

Additionally we have isolated the $HGPRT_{Illinois}$ cDNA using PCR amplification and cloned it into a eukaryotic expression vector. The 13 bp deletion has been confirmed by sequencing of the plasmid construct. The resulting plasmid has been used to transfect human fibroblasts known to be devoid of detectable

HGPRT mRNA. Selection of transfectants in hypoxanthine, aminopterin, thymidine (HAT) medium has allowed us to isolate clones expressing low levels of HGPRT activity [16]. We are currently expanding the clones transfected with the HGPRT$_{Illinois}$ plasmid construct for functional HGPRT analysis.

The results of these studies suggest than the *GTG* codon present in HGPRT$_{Illinois}$ can function inefficiently as an initiation codon, illustrated as follows:

$$\begin{array}{llccc}
 & & \text{(Val)} & \text{Ala} & \text{Thr} \\
\text{HGPRT}_{\text{Illinois}} & 5'...\text{CCCGCGCGCC} & GTG & \text{GCG} & \text{ACC}...3'
\end{array}$$

This in turn allows the production of extremely low levels of HGPRT protein that is normal in size, immunoreactivity, and catalytic function and produces a partially deficient phenotype in the affected individual. In contrast, alternative translational initiation does not appear to occur in the *ATA* codon of HGPRT$_{1151}$, no normal HGPRT protein is produced, and a Lesch-Nyhan phenotype results.

Previously, alternative initiator codon usage has been described primarily in viruses and lower eukaryotes [17–23]. However, the use of non-ATG initiation codons has rarely been identified in humans [24]. Precedence also exists for the use of alternative initiator codons in the production of purine biosynthetic enzymes. The autosomal locus of PRPP synthetase (PRPS3) has been shown to initiate translation from a ACG initiation codon [25]. The use of the GTG codon has been described in the *Drosophila* choline acetyltransferase gene [20]. Of the six possible alternative initiation codons previously studied, GTG was shown to be the most efficient, producing 3%–5% of the amount of protein produced by an ATG initiation codon [26].

The possibility that translation initiation from a GTG codon occurs in HGPRT$_{Illinois}$ is supported by considering the genetic context of the codon [27–30]. Kozak's studies with preproinsulin have shown that initiation is most efficient in eukaryotes when the start codon is in the context GC(A/G)CC*ATG* G. The deletion in HGPRT$_{Illinois}$ recapitulates this consensus sequence and replaces the ATG with GTG.

In conclusion, the genetic analysis of HGPRT deficiency states has already provided us with valuable information regarding the structure-function aspects of normal and abnormal HGPRT catalytic activity. The continued genetic analysis of HGPRT deficiency states will undoubtedly provide us with additional insight into general mechanisms of human biology, including replication, re-combination, transcription, translation, and mutagenesis. The use of HGPRT deficiency as a model system for eukaryotic gene therapy will also continue to be an area of active and productive research.

# References


1. Lesch M, Nyhan WL (1964) A familial disorder of uric acid metabolism and central nervous system function. Am J Med 36: 561–570
2. Kelley WN, Rosenbloom FM, Henderson JF, Seegmiller JE (1967) A specific enzyme defect in gout associated with overproduction of uric acid. Proc Natl Acad Sci 57: 1735–1739

3.  Ogasawara N, Stout JT, Goto H, Sonta S, Matsumoto A, Caskey CT (1989) Molecular analysis of a female Lesch-Nyhan patient. J Clin Invest 84: 1024–1027

4.  Davidson BL, Palella TD, Kelly WN (1988) Human hypoxanthine-guanine phosphoribosyl transferase: a single nucleotide substitution in cDNA clone isolated from a patient with Lesch-Nyhan syndrome. Gene 68: 85–91

5.  Wilson JM, Stout JT, Palella TD, Davidson BL, Kelley WN, Caskey CT (1986) A molecular survey of hypoxanthine-guanine phosphoribosyl-transferase deficiency in man. J Clin Invest 77: 188–195

6.  Gibbs RA, Nguyen P, McBride LJ, Koepf SM, Caskey CT (1989) Identification of mutations leading to Lesch-Nyhan syndrome by automated direct DNA sequencing of in vitro amplified cDNA. Proc Natl Acad Sci 86: 1919–1923

7.  Tarle SA, Davidson BL, Wu VC, Zidar FJ, Seegmiller JE, Kelley WN, Palella TD (1991) Determination of the mutations responsible for the Lesch-Nyhan syndrome in 17 subjects. Genomics 10: 499–501

8.  Davidson BL, Pashmforoush M, Kelley WN, Palella TD (1989) Human hypoxanthine-guanine phosphoribosyltransferase deficiency. J Biol Chem 264: 520–525

9.  Davidson BL, Tarle SA, van Antwerp M, Gibbs DA, Watts RWE, Kelley WN, Palella TD (1991) Identification of 17 independent mutations responsible for human hypoxanthine-guanine phosphoribosyltransferase (HGPRT) deficiency. Am J Hum Genet 48: 951–958

10. Yang TP, Stout JT, Konecki DS, Patel PI, Alford RL, Caskey CT (1988) Spontaneous reversion of novel Lesch-Nyhan mutation by HGPRT gene rearrangement. Somat Cell Mol Genet 14: 293–303

11. Saiki RK, Gelfand DH, Stoffel S, Scharf SJ, et al (1988) Primer directed enzymatic amplification of DNA with thermostable DNA polymerase. Science 239: 487–491

12. Goodenow M, Huet T, Saurin W, Kwok S, Wain-Hobson S (1989) HIV1 isolates are rapidly evolving quasispecies: evidence for viral mixtures and preferred nucleotide substitutions. J Acquir Immune Def Synd 2: 344–352

13. Chou PY, Fasman GD (1978) Empirical predictions of protein conformation. Ann Rev Biochem 47: 251–276

14. Hopp TP, Woods KR (1981) Prediction of protein antigenic determinants from amino acid sequences. Proc Natl Acad Sci USA 78: 3824–3828

15. Monnat RJ Jr., Chiaverotti TM, Hackmann AFM (1991) Molecular analysis of human HGPRT gene deletions and duplications. Adv Exp Med Biol 309 B: 113–116

16. Kennett RH (1979) Cell fusion, in Methods in Enzymology, Vol 52, Colowick SP, Kaplan NO eds. pp 345–359 Academic Press, New York

17. Becerra SP, Rose JA, Hardy M, Baroudy BM, Anderson CW (1985) Direct mapping of adeno-associated virus capsid proteins B and C: a possible ACG initiation codon. Proc Natl Acad Sci USA 82: 7919–7923

18. Curran J, and Kolakofsky D (1988) Ribosomal initiation from an ACG codon in the Sendai virus P/C mRNA. Embo Journal 7: 245–251

19. Acland P, Dixon M, Peters G, Dickson C (1990) Subcellular fate of the Int-2 oncoprotein is determined by choice of initiation codon Nature 343: 662–665

20. Sugihara H, Andrisani V, Salvaterra PM (1990) *Drosophila* choline acetyltransferase uses a non-AUG initiation codon and full length RNA is inefficiently translated. J Biol Chem 265: 21714–21719

21. Lemair P, Vesque C, Schmitt J, Stunnenberg H, Frank R, Charnay P (1990) The serum-inducible mouse gene Krox-24 encodes a sequence specific transcriptional activator. Mol Cell Biol 10: 3456–3467

22. Prats A, Wang G, Darlix J (1989) CUG initiation codon used for the synthesis of a cell surface antigen coded by the murine leukemia virus. J Mol Biol 205: 3633–372
23. Peabody DS (1987) Translation initiation at an ACG triplet in mammalian cells. J Biol Chem 262: 11847–11851
24. Florkiewicz RZ, Sommer A (1989) Human basic fibroblast growth factor gene encodes four polypeptides: three initiate translation from non-AUG codons. Proc Natl Acad Sci USA 86: 3978–3981
25. Taira M, Iizasa T, Shimada H, Kudoh J, Shimuzu N, Tatibana M (1990) A human testis-specific mRNA for phosphoribosylpyrophosphate synthetase that initiates from a non-AUG codon. J Biol Chem 265: 16491–16497
26. Prats H (1989) High molecular mass forms of basic fibroblast growth factor are initiated by alternative CUG codons. Proc Natl Acad Sci USA 80: 1836–1840
27. Kozak M (1986) Point mutations define a sequence flanking the AUG initiator codon that modulates translation by eukaryotic ribosomes. Cell 44: 283–292
28. Kozak M (1987) At least six nucleotides precending the AUG initiator codon enhance translation in mammalian cells. J Mol Biol 196: 947–950
29. Kozak M (1989) Context effects and inefficient initiation at non-AUG codons in eukaryotic cell-free translation systems. Mol Cell Biol 9: 5073–5080
30. Kozak M (1990) Downstream secondary structure facilitates recognition of initiator codons by eukaryotic ribosomes. Proc Natl Acad Sci USA 87: 8301–8305

# 1B  Adenine Phosphoribosyltransferase (APRT) Deficiency

# 1  The Clinical Aspects of APRT Deficiency

U. GRESSER and B. S. GATHOF

## Introduction

Adenine phosphoribosyltransferase (APRT) deficiency was first detected and described by Kelley et al. in 1968. The patient, presenting with a disorder of the lipoprotein metabolism, was planned to participate as one of the control series in a study of HGPRT and APRT activities in patients with gouty arthritis and nephrolithiasis. His APRT activity was reduced to about 25% of the normal value; this was also the case in his two daughters and his mother. All of them were clinically asymptomatic. The first two patients with complete APRT deficiency were reported by Cartier and Hamet in 1974 and by Simmonds et al. in 1976. Excretion of the purine metabolite 2,8-dihydroxyadenine (2,8-DHA) was detected as the source of crystalluria and nephrolithiasis in these patients.

Since then several cases of complete and partial APRT deficiency have been described in different populations (review in Simmonds et al. 1989; see also Abe et al. 1987; Glicklich et al. 1988; Hesse et al. 1988; Hönecke and Butz 1989; Leusmann and Schmidt 1990; Takeuchi et al. 1989; Usenius et al. 1988; Zöllner and Gresser 1990). Determination of APRT activities in red cell lysates led to a differentiation (Table 1) between the Caucasian (or type I) APRT deficiency associated with the *APRT*Q0* allele and the Japanese (or type II) APRT deficiency related to the *APRT*J* allele. The normal allele was named *APRT*1*.

**Table 1.** Genotypes and phenotypes in APRT deficiency (modified from Kamatani et al. 1989)

| Genotype | Approximate enzyme activity (% of normal) | 2,8-DHA lithiasis | Sensitivity to adenine analogues |
|---|---|---|---|
| *APRT*1/APRT*1* | 100 | – | sensitive |
| *APRT*1/APRT*Q0* | 25 | – | sensitive |
| *APRT*Q0/APRT*Q0* | 0 | + | resistant |
| *APRT*1/APRT*J* | 50 | – | sensitive |
| *APRT*J/APRT*J* | 25 | + | resistant |
| *APRT*J/APRT*Q0* | 25 | + | resistant |

*APRT*1*   = common (normal) allele;
*APRT*Q0* = Caucasin (type I) mutant allele;
*APRT*J*   = Japanese (type II) mutant allele.

## Epidemiology

Several studies on enzyme activities, most of them in patient populations show that heterozygotes (genotype *APRT*1/APRT*Q0*) are detected with a prevalence of 0.4%–1.1% (Fox 1977; Johnson et al. 1977). According to this a frequency of 1:40.000 to 1:250.000 for homozygotes (genotype *APRT*Q0/ APRT*Q0*) can be calculated. Prevalences in the same range have been reported from Japan for the type II deficiency (Kamatani et al. 1987). APRT deficiency is inherited in an autosomal recessive manner; the sex distribution of the homozygous male and female cases is 1:1.

More than 40 cases of complete APRT deficiency (type I) have been reported in the literature. Twelve of them are Japanese. Two compound heterozygous patients with the *APRT*J/Q0* allele have been described, and in a series of 51 Japanese families with APRT deficiency, five patients (10%) are suspected to be compound heterozygous (Kamatani et al. 1990).

## Clinical Symptoms

Heterozygotes with APRT deficiency are clinically asymptomatic. About 10% of the Caucasian (Cartier et al. 1980; Chevet et al. 1984; Gault et al. 1981; van Acker et al. 1977) and Japanese type I homozygotes published are clinical asymptomatic. The majority, about 85%–90% (Simmonds et al. 1989), of homozygous patients present with symptoms caused by crystalluria and nephrolithiasis such as discharge of stones with and without renal colics, acute renal failure or chronic renal failure. According to the literature about 15% of the homozygous patients with nephrolithiasis have to undergo dialysis and on some patients kidney transplantation has been performed. Some cases of APRT deficiency have been diagnosed in early childhood. Others have been detected in adults with recurrent nephrolithiasis, even once in a patient who had already undergone kidney transplantation (Glicklich et al. 1988). The same clinical symptoms are observed in patients with types I and II APRT deficiency, though type II individuals tend to develop symptoms at a higher frequency (Kamatani et al. 1989). The clinical symptoms and course in APRT deficiency are illustrated by some cases summarized in Table 2 and by the following case report.

*Case report.*    Homozygous twin brothers [originally described by Zöllner and Gresser (1990) and Leusmann and Schmidt (1990)] presented to our outpatient department because of recurrent nephrolithiasis at the age of 13 years. Birth and early childhood were clinically inconspicuous. At the age of 6 years both brothers were examined at a children's hospital because of headaches. Abdominal ultrasound revealed nephrolithiasis in both boys. At the age of 10 years both boys started to discharge stones and began to suffer from renal colics. Laboratory analysis detected "uric acid". For therapy the boys were advised to drink large quantities. At the age of 13 years the stones were diagnosed to be

2,8-DHA stones (Leusmann and Schmidt 1990). Complete APRT deficiency was found in both boys (laboratory Y.S. Shin, Munich). Examination of the family showed partial APRT deficiency in both parents and the older brother of the twins. Allopurinol was administered for a short time and discontinued because of headaches and skin papules. Subsequently one boy suffered from impaction of a renal stone, underwent lithotripsy and was referred to our hospital. Since then we have treated the boys with low purine diet, high fluid intake and allopurinol. The excretion of 2,8-DHA, renal colics and discharge of stones have decreased, but nephrolithiasis detectable on ultrasound (Fig. 1) has persisted. Molecular genetic analysis of the family revealed a T insertion (Gathof et al. 1991) which leads to aberrant splicing. This mutation was also demonstrated in other family members (Gathof and Zöllner 1992) using a new restriction enzyme, *Mse*I, which recognizes the mutation but not the normal APRT gene.

**Table 2.** Symptoms and course in selected cases of APRT deficiency

| Publication | Sex | Age at diagnosis (years) | APRT activity (% of normal) | Symptoms and course |
| --- | --- | --- | --- | --- |
| Cartier and Hamet 1974 | M | 4 | < 0.01 | At 2 dysuria, at 2 ¹/₂ y 2,8-DHA nephrolithiasis |
| van Acker et al. 1977 | M | 3 | < 0.1 | Urinary gravel since birth, urolithiasis at 1 ²/₃ y., asymptomatic with allopurinol, asymptomatic brother |
| | M | 7 | < 0.1 | |
| Greenwood et al. 1982 | F | 4 | < 0.1 | At 2 abdominal pain and hematuria, at 4 acute renal failure with coma and anuria requiring dialysis (under macrobiotic high-purine diet), persistently impaired renal function with low-purine diet and allopurinol treatment |
| Glicklich et al. 1988 | F | 42 | ? | Since age 24 recurrent nephrolithiasis, progressive renal failure requiring dialysis and renal transplantation, nephrolithiasis in the engrafted kidney, impaired renal function, free of symptoms with predisone, azathioprine, allopurinol, furosemide |

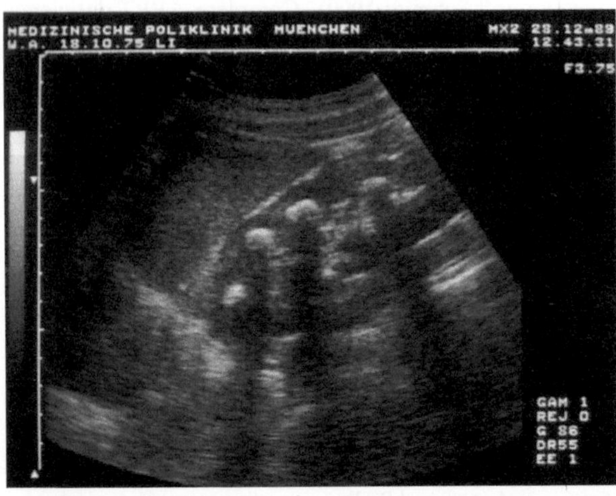

**Fig. 1.** Ultrasonogram of the left kidney of one of the twins with complete APRT deficiency at the age of 14 years. Multiple hyperechoic echo patterns with acoustic shadowing characteristic for concrements

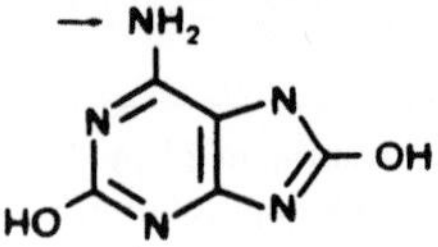

2,8-DIHYDROXYADENINE

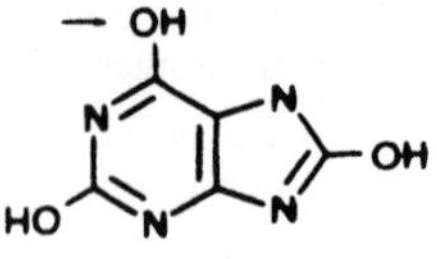

URIC ACID

**Fig. 2.** Molecular structure of 2,8-dihydroxyadenine and uric acid (*arrows* indicate the difference in structure)

In addition to nephrolithiasis several different symptoms, for instance juvenile gout, have been described in patients with APRT deficiency (Delbarre et al. 1974; Emmerson et al. 1975; Fox 1977), but no association has been confirmed.

## Diagnosis

As illustrated by the case described correct identification of the renal stones is most important in the diagnosis of APRT deficiency. 2,8-DHA stones are misdiagnosed as uric acid stones by standard laboratory techniques, because 2,8-DHA and uric acid differ in one position only (Fig. 2). Like uric acid stones 2,8-DHA stones are radiolucent. 2,8-DHA stones, however, are pale and friable, whereas uric acid stones are yellowish, hard and difficult to crush. 2,8-DHA stones are resistant to uricase. They can be detected by virtue of a characteristic pattern on infrared spectrophotometry, mass spectrometry and X-ray crystallography, by their UV spectrum in both acid and alkali and by high-pressure liquid chromotography (Simmonds et al. 1989).

## Treatment and Prognosis

Dietary purine restriction and high fluid intake are the basic treatment in asymptomatic and symptomatic homozygotes. Additionally the xanthine oxidase inhibitor allopurinol should be given in stone-formers. The dose should be adapted to the individual needs and risks of the patient. In case of impaired renal function allopurinol should be reduced. Alkali is not recommended as it does not increase the solubility of 2,8-DHA (Table 3) and has led to increased stone formation in combination with allopurinol in some cases. Consequent therapy as described can diminish and eliminate the excretion of 2,8-DHA.

**Table 3.** Solubility of different purines in human urine at two different urinary pH levels (Simmonds et al. 1989)

| Purine | ph 5 | | ph 8 | |
| --- | --- | --- | --- | --- |
| | mg/dl | mmol/l | mg/dl | mmol/l |
| Uric acid | 15 | 0.9 | 200 | 12.0 |
| Xanthine | 8 | 0.5 | 13 | 0.9 |
| 2,8-DHA | 0.3 | 0.02 | 0.5 | 0.03 |

Untreated patients with complete APRT deficiency have daily renal 2,8-DHA excretion of about 50–80 mg. Solubility of 2,8-DHA in urine: maximum 5 mg/l.

One of the questions still open is the contrast between the small number of homozygous cases reported in the literature and the frequency of heterozygotes. In addition to the heterogeneity of clinical expression, one reason may be misinterpretation of 2,8-DHA stones as uric acid stones due to difficulties in laboratory analysis. Also, a possibly lethal mutation upstream from the APRT locus associated with complete APRT deficiency has been postulated (Tischfield et al. 1990).

## Conclusions

In case of radiolucent nephrolithiasis in children and adults, as described here, the diagnosis of APRT deficiency should be considered. APRT deficiency requires early and consequent therapy to prevent renal damage.

# References


Abe S, Hayasaka K, Narisawa K (1987) Partial and complete adenine phosphoribo-syltransferase deficiency associated with 2,8-dihydroxyadenine urolithiasis: kinetic and immunochemical properties of APRT. Enzyme 37: 182–188

Cartier P, Hamet M (1974) Une nouvelle maladie métabolique: le déficit complet en adénine-phosphoribosyltransférase avec lithiase de 2,8-dihydroxyadénine. C R Acad Sci Paris 279: 883–886

Cartier P, Hamet M, Vincens A, Perignon JL (1980) Complete adenine phosphoribosyl-transferase (APRT) deficiency in two siblings: report of a new case. Adv Exp Med Biol 112 A: 343–348

Chevet P, le Pogamp SG, Gary J, Daudon M, Hamet M (1984) 2,8-Dihydroxyadenine (2,8-DHA) urolithiasis in an adult: complete adenine phosphoribosyltransferase (APRT) deficiency. Family study. Kidney Int 26: 226

Delbarre F, Auscher C, Amor B (1974) Gout with adenine phosphoribosyltransferase deficiency. Biomed 21: 82–85

Emmerson BT, Gordon RB, Thompson L (1975) Adenine phosphoribosyltransferase de-ficiency: its inheritance and occurrence in a female with gout and renal disease. Aust NZ J Med 5: 440–446

Fox IH (1977) Purine enzyme abnormalities: a four year experience. Adv Exp Med Biol 76 A: 265–269

Gathof BS, Sahota A, Gresser U, Chen J, Stambrook PJ, Tischfield JA, Zöllner N (1991) Identification of a splice mutation at the adenine phosphoribosyltransferase locus in a German family. Klin Wochenschr 69: 1152–1155

Gathof BS, Zöllner N (1992) A new restriction enzyme MseI applied for the detection of a possibly common mutation of the APRT locus. Clin Investig 70: 535

Gault MH, Simmonds HA, Snedden W, Dow D, Churchill DN, Penney H (1981) Uroli-thiasis due to 2,8-dihydroxyadenine in an adult. N Engl J Med 305: 1570–1572

Glicklich D, Gruber HE, Matas AJ, Tellis VA, Gattu K, Finley K, Salem C, Soberman R, Seegmiller JE (1988) 2,8-Dihydroxyadenine urolithiasis: report of a case first diagnosed after renal transplant. Q J Med 69: 785–793

Greenwood MC, Dillon MJ, Simmonds HA, Barratt TM, Pincott JR, Metreweli C (1982) Renal failure due to 2,8-dihydroxyadenine urolithiasis. Eur J Pediatr 138: 346–349

Hesse A, Miersch WD, Classen A, Thon A, Doppler W (1988) 2,8-Dihydroxyadeninuria: laboratory diagnosis and therapy control. Urol Int 43: 174–178

Hönecke K, Butz M (1989) 2,8-Dihydroxyadeninstein: die Bedeutung der exakten Stein-analyse. Urologe 28: 361–362

Johnson LA, Gordon RB, Emmerson BT (1977) Adenine phosphoribosyltransferase – a simple spectrophotometric assay and the incidence of mutations in the normal popula-tion. Bichem Genet 15: 265–272

Kamatani N, Terai C, Kuroshima S, Nihioka K, Mikanagi K (1987) Genetic and clinical studies on 19 families with adenine phosphoribosyltransferase deficiencies. Hum Genet 75: 163–168

Kamatani N, Kuroshima S, Terai C, Hakoda M, Nishioka K, Mikanagi K (1989) Diagnosis of genotypes for adenine phosphoribosyltransferase (APRT) deficiency. Adv Exp Med Biol 253 A: 51–58

Kamatani N, Kuroshima S, Yamanaka H, Nakashe S, Take J, Hakoda M (1990) Identification of a compound heterozygote for adenine phosphoribosyltransferase deficiency (*APRT*J*/ *APRT*Q0*) leading to 2,8-dihydroxyadenine lithiasis. Hum Genet 85: 500–504

Kelley WN, Levy RI, Rosenbloom FM, Henderson JF, Seegmiller JE (1968) Adenine phosphoribosyltransferase deficiency: a previously undescribed genetic defect in man. J Clin Invest 47: 2281–2289

Leusmann DB, Schmidt G (1990) Urolithiasis bei 2,8-Dihydroxyadeninurie: Vorstellung von drei weiteren Fällen. Z Urol Nephrol 83: 383–389

Simmonds HA, van Acker KJ, Cameron JS, Snedden W (1976) The identification of 2,8-dihydroxyadenine, a new component of urinary stones. Biochem J 157: 485–487

Simmonds HA, Sahota AS, van Acker KJ (1989) Adenine phosphoribosyltransferase deficiency and 2,8-dihydroxyadenine lithiasis. In: Scriver CR, Beaudet AL, Sly WS, Walle D (eds) The metabolic basis of inherited disease, 6th edn. McGraw-Hill, New York, pp 1029–1044

Takeuchi F, Kamatani N, Nishida Y, Miyamoto T (1989) Erythrocyte adenine PRPP availability in two types of APRT deficiency using silicone oil method. Adv Exp Med Biol 253 A: 35–41

Tischfield JA, Chen J, Stambrook PJ (1990) Whither the APRT? Abstract volume 16, molecular biology of APRT conference, Indiana University, Bloomington

Usenius JP, Roupuro ML, Usenius R (1988) Adenine phosphoribosyltransferase urolithiasis in a 48 year old woman. Br J Urol 62: 521–524

Van Acker KJ, Simmonds HA, Potter C, Cameron S (1977) Complete deficiency of adenine phosphoribosyltransferase. N Engl J Med 297: 127–132

Zöllner N, Gresser U (1990) Nephrolithiasis in twins with APRT deficency. Stones as a marker of an inborn error of metabolism. Bildgebung/Imaging 57: 64–66

# 2   The Biochemical Basis of APRT Deficiency

Y. S. SHIN

## Abstract

We found adenine phosphoribosyltranferase (APRT) deficiency in two children and two adults of two unrelated German families and one Arabic family. All patients exhibited 2,8-dihydroxyadenine (2,8-DHA) urinary track stone and APRT activity was found to be totally absent in red blood cells (normal range: 0.7–1.5 µmol/min/g Hb). In addition to the twins described in the previous article by Gresser and Gathoff, a 25-year-old female with a radiolucent 2,8-DHA stone in the right kidney and a 49-year-old Arabic male with scarring of the kidney were found to be suffering from APRT deficiency. All were treated with allopurinol and the urinary tract crystals disappeared. Eight members of two German families were found to be heterozygous for APRT deficiency with the APRT activity in red blood cells 20%–30% of the normal values (0.2–0.35 µmol/min/g Hb). The heterozygotes are all healthy and without any clinical symptoms. Although the clinical symptoms of homozygotes and heterozygotes are quite variable, early detection of the disease and the subsequent initiation of therapy seem to be essential to preventing kidney damage.

## Introduction

A deficiency of adenine phosphoribosyltransferase (EC 2.4.2.7; APRT), an autosomal recessive trait, is characterized by 2,8-dihydroxyadenine (2,8-DHA) urolithiasis (Simmonds et al. 1989; Cartier and Hammet 1974). APRT is a purine salvage enzyme which catalyzes the conversion of adenine to adenylic acid (5'-AMP) in the presence of 5-phosphoribosyl-1-pyrophosphate (Fig. 1). In subjects with APRT deficiency adenine is not salvaged to AMP but oxidized to 8-hydroxy-adenine (8-HA) and to 2,8-DHA by xanthine oxidase (Fig. 1). These relatively insoluble products lead to the formation of urinary tract stones in most patients with APRT deficiency (Simmonds et al. 1989). Two types of the defect have been

*Abbreviations:* APRT adenine phosphoribosyltransferase, HGPRT hypoxanthine guanine phosphoribosyltransferase, AMP adenosine monophosphate, IMP inosine monophosphate, PRPP 5'-phosphoribosyl-1-pyrophosphate, 8-HA 8-hydroxyadenine 2,8-DHA, 2,8-dihydroxyadenine

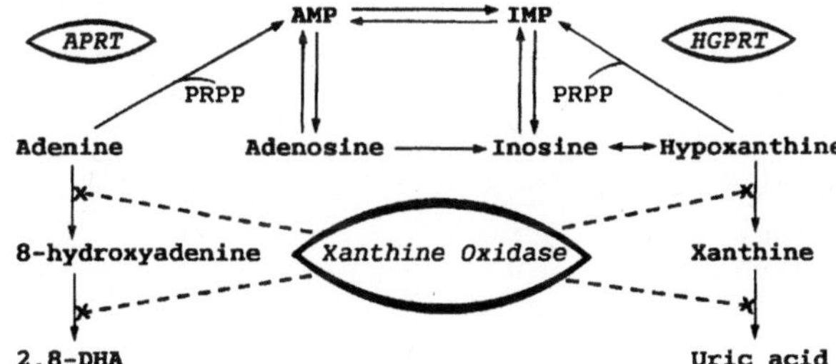


**Fig. 1.** Purine salvage pathway and the xanthine oxidase reaction

**Table 1.** Adenine phosphoribosyltransferase deficiency

| Subject | Residual RBC activity (%) | Estimated frequency | Clinical symptoms |
|---|---|---|---|
| Complete deficiency homozygote | 0 | 1:40000 − 100000 | 2,8-DHA stone Hematuria Infection Kidney damage |
| Heterozygote | 20–30 | 1:100–250 | None |
| Partial deficiency Homozygote | | ? Only Japanese | Urolithiasis |
| Heterozygote | 49–66 | Only Japanese | None |

described (Table 1), complete and partial deficiency according to the degree of the residual APRT activity in red blood cells. The high frequency of the heterozygosity in Caucasian populations (0.4%–1.1%) suggests that the disease may be more prevalent than reported so far. The clinical manifestations are also variable from symptomless to colic, hematouria and urinary tract infection. We report here biochemical basis of four patients with complete APRT deficiency from two German and one Arabic families.

## 2,8-Dihydroxyadenine

Patients with complete APRT deficiency excrete three adenine derivatives in urine, mostly adenine and 2,8-DHA and a small amount of 8-HA. All other urinary purine products such as uric acid and adenine metabolites, which amount to 20%–30% of total purine excretion, are reported to be in the normal range even on a low purine diet (Simmonds et al. 1989). 2,8-DHA which is secreted by the kidney is extremely insoluble (Table 2). Even though there are many similarities in the macroscopic appearances of uric acid and 2,8-DHA stones (Tables 3 and 4), there are various ways to differentiate them. When a

**Table 2.** Solubility of purine in human urine

| Purine | pH 5.0 (mg/dl) | pH 8.0 (mg/dl) |
|---|---|---|
| Uric acid | 15.0 | 200 |
| Xanthine | 8.0 | 13 |
| 2,8-DHA | 0.3 | 0.5 |

**Table 3.** Characteristics of purine stones

| | Uric acid | Xanthine | 2,8-DHA |
|---|---|---|---|
| Color | Yellowish | Orange, brown | Pale grey |
| Form | Round | Oval | Round |
| Surface | Smooth | Smooth | Rough |
| Consistency | Hard | Laminated | Crushable |

**Table 4.** Differentiation of 2,8-DHA stones from uric acid stones

| Similar | Dissimilar |
|---|---|
| Colorimetric analysis | Resistance to uricase |
| phosphomolybdate, murexide reaction) | Infrared spectrometry |
| Thermogravimetric analysis | Mass spectrometry |
| Alkaline UV spectrum | X-ray crystallography |
| X-ray (radiolucent) | HPLC retention time |
| | Acid UV spectrum |

patient has uric acid-like stones, renal colic and/or ultrasound detectable scars in the kidneys but has a normal serum uric acid concentration and no gouty attacks, one could suspect 2,8-DHA lithiasis.

## Adenine Phosphoribosyltransferase

APRT (EC 2.4.2.7) catalyzes the formation of 5'-AMP and pyrophosphate from adenine and phosphoribosylpyrophosphate (PRPP). APRT, a purine salvage enzyme, can be assayed by various methods (Arnold and Kelly 1978; Fairbanks et al. 1983). In our laboratory a radioisotopic method using [$^{14}$C]adenine has been used (Shin-Bühring et al. 1980). The separation of radioactive 5'-AMP from adenine was carried out by using a mini-DEAE-cellulose column, a system suitable for multiple analyses.

    The APRT activity in red cell lysates, as shown in Table 5, is about one third of the HGPRT activity (Shin-Bühring et al. 1980). The APRT values between normals and heterozygotes (about 20%–30%) as well as between homozygotes and heterozygotes (Table 5 and Fig. 2) are, however readily distinguishable. The pedigree of the twins described in the previous article well documents the autosomal recessive inheritance pattern of APRT deficiency (Fig. 2). In Table 6, the distribution of both APRT and hypoxanthine guanine phosphoribosyltransferase (HGPRT) in various human tissues has been summarized. In contrast to the activity in red blood cells, APRT activity in human tissues was similar to that of HGPRT, except perhaps in the liver.

Various genetic APRT deficiency mutants (Kamotani et al. 1987) have been described, indicating the existence of mutant isoenzymes. Mutant APRT enzymes from homozygous patients (Japanese-type partial APRT deficiency) were reported to have different kinetic properties such as affinity for PRPP or a stabilizing effect of PRPP (Fujimori et al. 1986). It is interesting to note the higher APRT activity in complete or partial deficiency of HGPRT (Kelly and Wyngarden 1983; Shin-Bühring et al. 1980), most probably due to increased PRPP levels which stabilize APRT (Greene et al. 1970). An early report also showed different red cell APRT isoenzyme pattern in a sample from a hyperuricemic boy by disc gel electrophoresis (Bakay and Nyhan 1971).

**Table 5.** Adenine phosphoribosyltransferase activity in red blood cells

| Subject | $n$ | μmol/min/g Hb |
|---|---|---|
| Deficiency | 4 | 0 |
| Heterozygote | 8 | 0.27 ± 0.06[a] <br> (0.20 − 0.35)[b] |
| Control | 71 | 1.05 ± 0.16[a] <br> (0.70 − 1.50)[b] |

[a] Mean ± SD.
[b] Range.

**Table 6.** Phosphoribosyltransferase activity in human tissues

| Sample[a] | $n$ | APRT[b] | HGPRT[b] |
|---|---|---|---|
| Fetal liver | 3 | 1.9–3.2 | 0.9–1.7 |
| Adult liver | 3 | 5.2–12.4 | 3.2–5.8 |
| Fetal intestine | 3 | 0.6–2.1 | 0.6–2.7 |
| Amniotic cells | 4 | 0.6–3.5 | 0.5–3.0 |
| Fibroblasts | 3 | 1.2–2.6 | 1.7–3.8 |
| Chorionic villi | 5 | 0.4–2.7 | 0.4–1.8 |

[a] Fetal livers and intestines were from 19–22 week-old fetuses; amniotic cells from amniotic fluids at 16–18 weeks of gestation; chorionic villi from biopsies at 8–12 weeks of gestation.
[b] nmol/mm/mg protein.

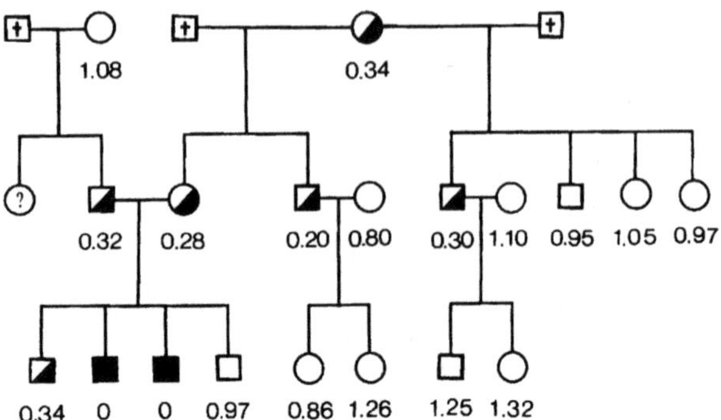

**Fig. 2.** Pedigree of a German family with complete APRT deficiency. ⊞, dead; □, normal; □ , heterozygotes; ■, propositus; [?] not studied. The *numbers* indicate APRT activity in red cell lysates in μmol/min/g Hb

## Prognosis and Treatment

Several studies have shown that adenine inhibits de novo purine synthesis (Wyngarden et al. 1959; Seegmiller et al. 1968). Other studies have, however, demonstrated that adenine actually increases uric acid levels by diminishing the PRPP pool (Marko et al. 1969; Schneeburger et al. 1981). This is in agreement with the ineffectiveness of adenine therapy in Lesch-Nyhan syndrome (Seegmiller et al. 1968). Endogenous adenine is now understood to originate by the polyamine pathway (Simmonds 1986) and is mostly secreted by the kidney followed by urinary excretion. Even though adenine metabolites in urine in APRT deficiency were not reduced by dietary purine restriction (Simmonds and van Acker 1983), a low purine diet is recommended, since exogenous adenine could aggravate the clinical symptoms (van Acker et al. 1977). Allopurinol, a potent xanthine oxidase inhibitor, decreases the amount of 2,8-DHA in the urine in most cases of APRT deficiency (Simmonds and van Acker 1983). In addition, a high fluid intake without alkali is also recommended. The prognosis of APRT deficiency is good, but completely depends on the state of renal function. Therefore, early diagnosis and subsequent initiation of therapy is important for prevention of permanent kidney damage.

*Acknowledgement.* I would like to express my sincere gratitude to Dr. U. Gresser, Med. Poliklinik, Munich, for preparing the pedigree shown in Fig. 2.

## References

Arnold WJ, Kelley WJ (1978) Adenine phosphoribosyltransferase. In: Hoffe PA, Jones ME (eds) Methods in Enzymology. Academic, New York, p 568

Bakay B, Nyhan WL (1971) The separation of adenine and hypoxanthine guanine phosphoribosyltransferase isoenzymes by disc gel electrophoresis. Biochem Gent 5: 81–90

Catier P, Hammet M (1974) Une nouvelle maladie métabolique: le déficit complet en adénine-phosphoribosyl transférase avec lithiase de 2,8-dihydroxyadénine. Comptes rendus hébdomadaires de séances de l'Académie des sciences. Série D, Sciences nouvelles 279: 883–886

Fairbanks LD, Goday A, Morris GS, Brolsma MFJ, Simmonds HA, Gibson T (1983) Rapid determination of purine enzyme activity in intact and lysed cells using high performance liquid chromatography with and without radiolabelled substrates. J Chrom 276: 427–431

Fujimori S, Akaoka I, Takeuchi N et al. (1986) Altered kinetic properties of a mutant adenine phosphoribosyltransferase. Metab 35: 187–192

Greene ML, Boyle JA, Seegmiller JE (1970) Substrate stabilization: genetically controlled reciprocal relationship of two human enzymes. Science 167: 887–889

Kamatani N, Kuroshima S, Terai C et al. (1987) Selection of human cells having two different types of mutations in individual cells (genetic/artificial mutants). Hum Genet 76: 148–152

Kelly WN, Wyngarden JB (1983) Clinical syndromes associated with hypoxanthine-guanine phosphoribosyltransferase deficiency. In: Stanbury JB, Wyngarden JB, Fredrickson DS, Goldstein JL, Brown MS (eds) The metabolic basis of inherited diseases, 5th edn. McGraw-Hill, New York, p 1115

Marko P, Gerlach E, Zimmer HG et al. (1969) Interrelationship between salvage pathway and synthesis de novo of adenine nucleotides in kidney slices. Hoppe-Seyler's Z Physiol Chem 41B: 671

Schneeburger W, Bach D, Hesse A, Valensieck W (1981) Circadian excretion of uric acid on a standard diet and purine load in Caoxalate stone formers and healthy controls. In: Smith LH, Robertson WG, Finlayson B (eds) Urilithiasis: clinical and basic research. Plenium, New York, p 51

Seegmiller JE, Klinenberg JR, Miller J, Watts RWE (1968) Suppression of glycine 15N incorporation into urinary uric acid by adenine-8-13C in normal and gouty subjects. J Clin Invest 47: 1193

Shin-Bühring YS, Osang M, Wirtz A et al. (1980) Prenatal diagnosis of Lesch-Nyhan syndrome and some characteristics of hypoxanthine-guanine phosphoribosyltransferase and adenine phosphoribosyltransferase in human tissues and cultivated cells. Pediatr Res 14: 825–829

Simmonds HA, van Acker KJ (1983) Adenine phosphoribosyltransferase deficiency: 2,8-dihydroxyadenine lithiasis. In: Stanbury JB, Wyngarden JB, Frederickson DS, Goldstein JL, Brown MS (eds) Teh metabolic basis of inherited diseases, 5th edn. McGraw-Hill, New York, p 485

Simmonds HA (1986) 2,8-Dihydroxyadenine lithiasis-expidermology, pathogenesis and therapy. In: Verhandlungen der deutschen Gesellschaft für innere Medizin. Bergmann, Munich, p 503

Simmonds HA, Sahota, AS, van Acker K (1989) Adenine phosphoribosyltransferase deficiency and 2,8-dihydroxyadenine lithiasis. In: Scriver CR, Beaudet AL, Sly WS, Valle D (eds) The metabolic basis of inherited diseases, 6th edn. McGraw-Hill, New York, pp 1029–1044

Van Acker KJ, Simmonds HA, Potter CF, Cameron JS (1977) Complete deficiency of adenine phosphoribosyltransferase: report of a family. Engl J Med 297: 127

Wyngarden JB, Seegmiller JE, Laster L, Blair AE (1959) Utilization of hypoxanthine, adenine and 4-amino-5-imidazolecarboximide for uric acid synthesis in man. Metab 8: 445

# 3  The Genetic Basis of APRT Deficiency

A. SAHOTA, J. CHEN, P. J. STAMBROOK, and J. A. TISCHFIELD

## Abstract

We have determined the molecular basis of adenine phosphoribosyltransferase (APRT) deficiency in 27 families. The APRT gene from each patient was amplified by the polymerase chain reaction, subcloned into M13mp18, and sequenced completely. Selected regions of the amplified fragment were also sequenced directly. DNA samples were also analyzed by *Sph*I and *Taq*I digestions and Southern blotting. A total of 18 different mutations alleles were identified. The mutations were nonrandomly distributed and suggested the presence of hot spots. At least three mutations appeared to have common ancestral origins. The paucity of intron or exon polymorphisms in these patients confirmed our previous studies that the APRT gene is highly conserved.

## Introduction

Adenine phosphoribosyltransferase (APRT; EC 2.4.2.7) is an enzyme of purine metabolism that catalyzes the synthesis of AMP from adenine and 5-phospho-ribosyl-1-pyrophosphate (PRPP). In the absence of APRT, adenine is oxidized by xanthine oxidase to the highly insoluble derivative, 2,8-dihydroxyadenine (DHA). The accumulation of this compound in the kidney can result in urolithiasis and, in severe cases, renal failure (Simmonds et al. 1989). Although APRT deficiency (McKusick 102600) is a relatively rare inborn error of metabolism, measurement of erythrocyte enzyme activity in large population groups suggests that about 1% of individuals are putative APRT heterozygotes (Simmonds et al. 1989).

Two phenotypic variants of APRT deficiency have been recognized. Type I APRT deficiency, which results in the complete absence of enzyme activity, is due to a variety of null mutant alleles collectively designated *APRT*Q0* (Fujimori et al 1985; Hidaka et al. 1987). Type II deficiency, which results in the production of APRT with a reduced affinity for PRPP, is due to the missense mutant allele *APRT*J* (Kamatani et al. 1987; Hidaka et al. 1988). Type I APRT deficiency has been found in many different ethnic groups, but type II deficiency has been found only in the Japanese. The inheritance pattern for both types of deficiency is autosomal recessive.

Homozygous type II APRT deficiency is due to a T→ C substitution in codon 136 (Met → Thr) of the APRT gene (Hidaka et al. 1988). A small percentage of type II patients are compound heterozygotes, with one *APRT*J* allele and one *APRT*Q0* allele (Kamatani et al. 1990a; Sahota et al. 1990). The molecular basis of type I deficiency is much more heterogeneous (Hidaka et al. 1987; Sahota et al. 1990; Chen et al. 1991a; Mimori et al 1991). Here we describe the identification of several additional *APRT*Q0* mutant alleles in type I patients. A part of this work has appeared in the proceedings of an earlier conference (Sahota et al. 1991).

## Patients

We have determined the molecular basis of APRT deficiency in 19 non-Japanese families (23 cases, including four siblings) from 14 countries. Three of the cases are unaffected heterozygotes. We have also studied eight Japaneses families, six with type II deficiency and two with type I deficiency. The national origin of the patients is shown in Table 1.

**Table 1.** Country of origin of patients with APRT deficiency

| Country | n | Comments |
|---|---|---|
| Australia | 2 | Sibs, heterozygotes |
| Austria | 1 | |
| Belgium | 2 | Sibs |
| Bermuda | 1 | |
| Britain | 1 | |
| Canada | 2 | Sibs |
| France | 1 | |
| Germany | 2 | Sibs |
| Greece | 1 | |
| Hungary | 1 | |
| Iceland | 5 | |
| Iraq | 1 | |
| Japan | 8 | 6 Type II, 2 type I |
| Pakistan | 1 | |
| USA | 2 | 1 Heterozygote |
| Total | 31 | |
| Excluding sibs | 27 | |

## Methods

Genomic DNA was isolated from blood or from cultured lymphoblasts and a 2.4 kilobase (kb) fragment containing the APRT gene was amplified by the polymerase chain reaction (PCR). The PCR product was subcloned into the bacteriophage M13mp18 and several clones from each individual were sequenced completely. Selected regions of the PCR-amplified DNA were also sequenced directly. DNA samples were also digested with *Sph*I and *Taq*I and analyzed by Southern blotting. Full details of the methodology have been described elsewhere (Chen et al. 1991a; Chen et al. 1991b; Sahota et al. 1991).

## Results

We have observed 18 different mutations (14 in non-Japanese and four in Japanese patients) in the 27 families we have studied (Tables 2 and 3). Some 50% of the mutations were located in exon 3. None of the patients had gross alterations in APRT, and no polymorphic amino acid substitutions were detected. We have noted two intron polymorphisms in some Japanese patients.

**Table 2.** Mutations at the APRT locus in Japanese patients

| $n$ | Base change | Amino acid change | Location | Mutation |
|---|---|---|---|---|
| 1 | A → G | $Met_1$ → Val | E 1 | Initiation codon |
| | GGCCCCA deletion | | E 3 | Frameshift after $Pro_{93}$ |
| 2 | T insertion | | E 2 | Frameshift after $Ile_{61}$ |
| 3 | T insertion | | E 2 | Frameshift after $Ile_{61}$ |
| | G → A | $Arg_{67}$ → Gln | E 3 | Missense |
| 4–8 | A → T | $Asp_{65}$ → Val | E 3 | Missense |
| 9 | A → T | $Asp_{65}$ → Val | E 3 | Missense |
| 10 | C → T | $Arg_{87}$ → Stop | E 3 | Nonsense |
| 11 | AC deletion | | E 3 | Frameshift after $Pro_{95}$ |
| 12[a] | T → C | $Leu_{110}$ → Pro | E 4 | Missense |
| 13 | A → T | $Ile_{112}$ → Phe | E 4 | Missense |
| | T → C | $Cys_{153}$ → Arg | E 5 | Missense |
| 14 | G → T | | I 4 | Splice donor |
| 15[a,b] | TTC deletion | $Phe_{173}$ | E 5 | Triplet deletion |
| | T insertion | | I 4 | Splice donor |
| 16 | T insertion | | I 4 | Splice donor |
| 17 | T insertion | | I 4 | Splice donor |
| 18 | T insertion | | I 4 | Splice donor |
| 19 | G → A | | E 3 | Missense |

E, exon; I, intron
[a] Siblings; [b] Originally described by Hidaka et al. (1987)

**Table 3.** Mutations at the APRT locus in Japanese patients

| $n$ | Change | Amino acid Location | | Mutation | |
|---|---|---|---|---|---|
| 1 | G → A | $\mathrm{Trp}_{98}$ → End | E 3 | | Nonsense |
|   | C → T | $\mathrm{Ala}_{99}$ → Ala | E 3 | | Neutral |
| 2 | G → A | $\mathrm{Trp}_{98}$ → End | E 3 | | Nonsense |
|   | C → T | $\mathrm{Ala}_{99}$ → Ala | E 3 | | Neutral |
|   | T → C | $\mathrm{Met}_{136}$ → Thr | E 5 | | Missense |
| 3–6 | T → C | $\mathrm{Met}_{136}$ → Thr | E 5 | | Missense |
| 7 | T → C | $\mathrm{Met}_{136}$ → Thr | E 5 | | Missense |
| 8 | CCGA insertion | | E 3 | | Frameshift after $\mathrm{Ile}_{86}$ |

E, exon.

Of the non-Japanese patients (Table 2), an insertion of a T nucleotide at the intron 4 splice donor site was found in four families, including the family originally described by Hidaka et al. (1987). The insertion results in an A → T change at the third base downstream from the splice site. A purine at this position appears to be essential for normal splicing (Carothers et al. 1990). The mutation leads to deletion of exon 4 from the mRNA because of aberrant splicing. A G → T transversion at this position was identified in another patient. This mutation is also expected to disrupt normal splicing. An insertion of a T nucleotide in exon 2 was found in two families. This insertion is expected to lead to a frameshift after $\mathrm{Ile}_{61}$. An A → T transversion in exon 3 ($\mathrm{Asp}_{65}$ → Val) was identified in six families (Chen et al. 1991a). The other non-Japanese patients had a variety of mutations, mainly single base changes and small deletions.

Of the Japanese patients (Table 3), one was homozygous for an exon 3 nonsense mutation ($\mathrm{Trp}_{98}$ → End) (Sahota et al. 1990). The same mutation was found in one allele from a type II compound heterozygote. The other allele from this patient had the *APRT*J* mutation. The *APRT*Q0* mutant allele from these two patients also contained a third position substitution at $\mathrm{Ala}_{99}$. Another type I patient had a four base pair insertion in exon 3. A patient (No. 7) with DHA lithiasis but with functional APRT activity in intact cells was shown to be a heterozygote, with one *APRT*J* allele and one wild-type allele (Sahota et al. 1991a). The remaining Japanese patients were homozygous for the *APRT*J* mutation.

## Discussion

A total of 46 families (51 cases, including 5 siblings) with type I APRT deficiency have been identified in 17 countries, excluding Japan (A Sahota and HA Simmonds, unpublished data). Approximately 80 families with APRT deficiency have been reported from Japan (Kamatani et al. 1988). Of these, about 70% have the type II defect and remainder have the type I defect (Kamatani et al. 1990b). A total of 18 different mutations have been observed (Tables 2 and 3) in the 27 families we have studied from 15 countries (Table 1). Twelve of the mutations were singles base pair changes (nine transitions and three transversions), three were small insertions (1–4 base pairs) and three were small deletions (2–7 base pairs).

Although the number of independent mutations we have found is relatively small, several striking features emerge from these data. There is a lack of major deletions or rearrangements among our sample. This is in contrast to a deficiency of the related purine enzyme, hypoxanthine-guanine phosphoribosyltransferase, in which major deletions and rearrangements account for about 15% of the mutations (Gibbs et al. 1989). Two mutations were located at the intron 4 splice donor site, and one of these was found in four unrelated families. This suggests that the intron 4 splice donor site is a hot spot for mutation. In principle, all splice donor sites within the gene ought to be equally mutable, yet there are no other splice donor site mutations within our sample. The middle of exon three also constitutes a mutationally hot region, since it contains seven of the nine mutations located in this exon.

An exon 3 missense mutation ($Asp_{65} \rightarrow Val$) was found in six families, five from Iceland and one from Britain (Chen et al. 1991a). It seems most likely that the mutation in all five Icelandic patients has a common ancestral origin, since most Icelanders are interrelated going back 12–14 generations. The absence of any sequence variations within the APRT gene or in the immediate flanking regions from these patients supports the notion of common ancestry. The finding of the same mutation in a British patient suggests that this mutation may have originated in mainland Europe. Previous studies have shown that the *APRT*J* mutation in patients with type II deficiency has a common ancestral origin (Kamatani et al. 1987; Kamatani et al 1990b). The nonsense mutation in exon 3 in Japanese patients with type I APRT deficiency also has a common ancestral origin (Sahota et al. 1990; Mimori et al. 1991).

No polymorphic amino acid substitutions were observed and only two intron polymorphisms were found. These data support our conclusion from earlier interspecific comparisons that the APRT gene is highly conserved (Broderick et al. 1987).

*Acknowledgement.*   We thank the physicians who referred patients to us, and the patients and their families for donating blood samples. This work was supported by NIH grant DK38185.

# References


Broderick TP, Schaff DA, Bertino AM, Dush MK, Tischfield JA, Stambrook PJ (1987) Comparative anatomy of the human APRT gene and enzyme: nucleotide sequence divergence and conservation of a nonrandom CpG dinucleotide arrangement. Proc Natl Acad Sci USA 84: 3349–3353

Carothers AM, Urlaub G, Mucha J, Harvey RG, Chasin LA, Grunberger D (1990) Splicing mutations in the CHO DHFR gene preferentially induced by (+/–) -3a,4b-dihydroxy-1a, 2a-epoxy-1,2,3,4-tetrahydrobenzo[c]phenanthrene. Proc Natl Acad Sci USA 87: 5464–5468

Chen J, Sahota A, Laxdal T, Scrine M, Bowman S, Cui C, Stambrook PJ, Tischfield JA (1991a) Identification of a single missense mutation in the adenine phosphoribosyltransferase gene from five Icelandic patients and a British patient. Am J Hum Genet 49: 251–255

Chen J, Sahota A, Stambrook PJ, Tischfield JA (1991b) Polymerase chain reaction amplification and sequence analysis of human mutant adenine phosphoribosyltransferase genes: the nature and frequency of errors caused by *Taq* DNA polymerase. Mutat Res 249: 169–176

Fujimori S, Akaoka I, Sakamoto K, Yamanaka H, Nishioka K, Kamatani N (1985) Common characteristics of mutant adenine phosphoribosyltransferase from four separate Japanese families with 2,8-dihydroxyadenine urolithiasis associated with partial enzyme deficiencies. Hum Genet 71: 171–176

Gathof BS, Sahota A, Gresser U, Chen J, Stambrook PJ, Tischfield JA, Zöllner N (1991) Identification of a splice mutation at the adenine phosphoribosyltransferase locus in a German family. Klin Wochenschr 69: 1152–1155

Gibbs RA, Nguyen P-N, McBride LJ, Koepf SM, Caskey CT (1989) Identification of mutations leading to the Lesch-Nyhan syndrome by automated direct DNA sequencing of in vitro amplified DNA. Proc Natl Acad Sci USA 86: 1919–1923

Hidaka Y, Palella TD, O'Toole TE, Tarle' SA, Kelley WN (1987) Human adenine phosphoribosyltransferase: identification of allelic mutations at the nucleotide level as a cause of complete deficiency of the enzyme. J Clin Invest 80: 1409–1415

Hidaka Y, Tarle' SA, Fujimori S, Kamatani N, Kelley WN, Palella TD (1988) Human adenine phosphoribosyltransferase deficiency: demonstration of a single mutant allele common to the Japanese. J Clin Invest 81: 945–950

Kamatani N, Terai C, Kuroshima S, Nishioka K, Mikanagi K (1987) Genetic and clinical studies on 19 families with adenine phosphoribosyltransferase deficiencies. Hum Genet 75: 163–168

Kamatani N, Sonoda T, Nishioka K (1988) Distribution of the patients with 2,8-dihydroxyadenine urolithiasis and adenine phosphoribosyltransferase deficiency in Japan. J Urol 140: 1470–1472

Kamatani N, Kuroshima S, Yamanaka H, Nakashe S, Take H, Hakoda M (1990a) Identification of a compound heterozygote for adenine phosphoribosyltransferase deficiency (*APRT*J*/*APRT*Q0*) leading to 2,8-dihydroxyadenine urolithiasis. Hum Genet 85: 500–504

Kamatani N, Kuroshima S, Hakoda M, Palella TD, Hidaka Y (1990b) Crossovers within a short DNA sequence indicate a long evolutionary history of the *APRT*J* mutation. Hum Genet 85: 600–604

Mimori A, Hidaka Y, Wu VC, Tarle' SA, Kamatani N, Kelley WN, Palella TD (1991) A mutant allele common to the Type I adenine phosphoribosyltransferase deficiency in Japanese subjects. Am J Hum Genet 48: 103–107

Sahota A, Chen J, Asako K, Takeuchi H, Stambrook PJ, Tischfield JA (1990) Identification of a common nonsense mutation in Japanese patients with Type I adenine phosphoribosyltransferase deficiency. Nuc Acids Res 18: 5915–5916
Sahota A, Chen J, Stambrook PJ, Tischfield JA (1991) Mutational basis of adenine phosphoribosyltransferase deficiency. Adv Expt Med Biol 309 B: 73–76
Sahota A, Chen J, Behzadian MA, Ravindra R, Takeuchi H, Stambrook PJ, Tischfield JA (1991a), 2,8-Dihydroxyadenine lithiasis in a Japanese patient heterozygous at the adenine phosphoribosyltransferase locus. Am J Hum Genet 48: 983–989
Simmonds HA, Sahota A, van Acker KJ (1989) Adenine phosphoribosyltransferase deficiency and 2,8-dihydroxyadenine lithiasis. In: Scriver CR, Beaudet AL, Sly WS, Valle D (eds) The metabolic basis of inherited disease, 6th edn. McGraw-Hill, New York, pp 1029–1044

# II  Hyperuricemia and Gout Caused by a Defect in Renal Transport

# 1 The Clinical Aspects of Hyperuricemia and Gout

U. GRESSER

Hyperuricemia is caused by a defect in renal transport in about 98% of all those affected. These patients have reduced uric acid clearance, which combined with purinrich food raises plasma uric acid levels to hyperuricemic values. Hyperuricemia is defined according to the limit of solubility of monosodium urate in extracellular fluid at pH 7.4 and 37°C to 6.5 mg/dl and more. In industrialized countries about 30% of the men and 3% of the women suffer from hyperuricemia [5, 6]. Depending on the duration and extent of the disease, patients develop gouty arthritis and nephrolithiasis (Table 1). In most patients with hyperuricemia caused by a defect in renal transport gouty arthritis occurs first. Nephrolithiasis is the first symptom in the majority of the patients who are overproducers because of an enzyme defect such as HGPRT deficiency. Some of the other diseases associated with hyperuricemia are hyperlipidemia, hypertension, obesity, atherosclerosis, hyperglycemia, diabetes mellitus, and ethanol consumption.

We distinguish between hyperuricemia without clinical symptoms, acute gouty attacks, and chronic gout. Patients with hyperuricemia may go without notable clinical symptoms for many years. Hyperuricemia nevertheless marks the beginning of chronic gout.

The typical case of acute gouty arthritis is an impressive and extremely painful monarthritis (Fig. 1). Without therapy it lasts for some days, yet with the help of colchicine, nonsteroidal antiinflammatory drugs, or corticosteroids it can be stopped within some hours (Table 2). Acute gouty attacks preferentially occur at the big toe, but they can happen at every joint.

**Table 1.** Prevalence of gouty arthritis and nephrolithiasis in relation to maximum serum uric acid levels (Framingham study, N = 2.283 men) [2]

| Serum uric acid level (mg/dl) | Gouty arthritis developed in (%) | Nephrolithiasis, typical history (%) |
| --- | --- | --- |
| 7.0 – 7.9 | 16.7 | 12.7 |
| 8.0 – 8.9 | 25.0 | 22.0 |
| ≥ 9.0 | 90.0 | 40.0 |

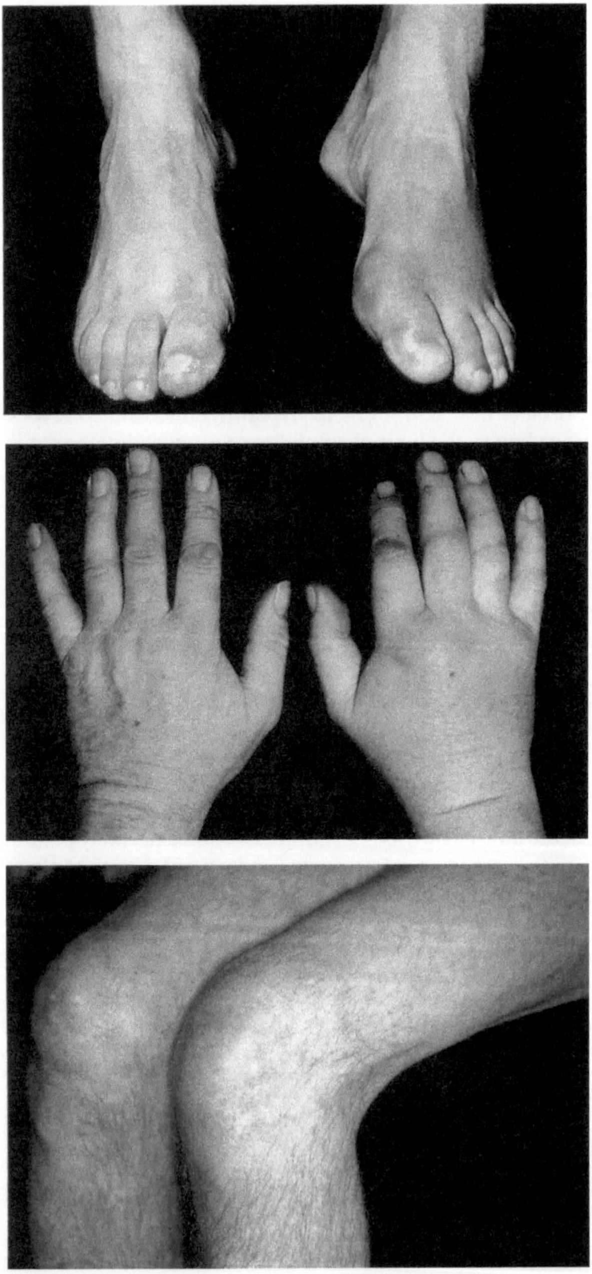

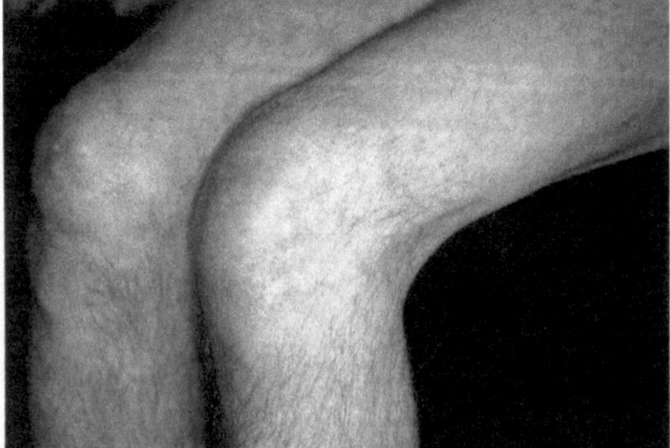

a

b

c

**Fig. 1a–c.** Acute gouty arthritis at the big toe of the left foot (**a**), at the MCP II of the right hand (**b**), and at the left knee (**c**) in different patients

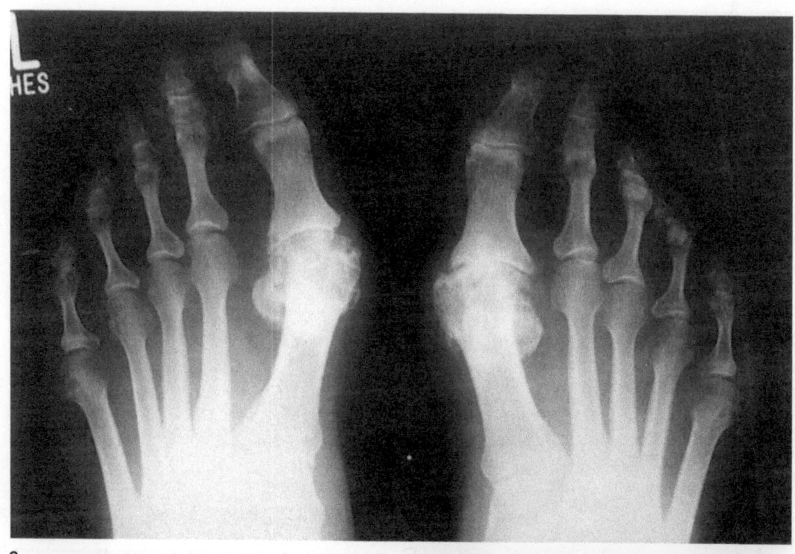
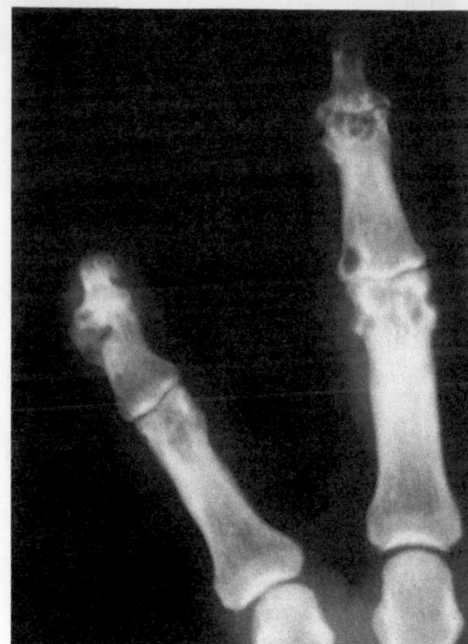

**Fig. 2a, b.** X-rays of feet (**a**) and fingers of the left hand (**b**) of a 58-year-old male with chronic tophaceous gout

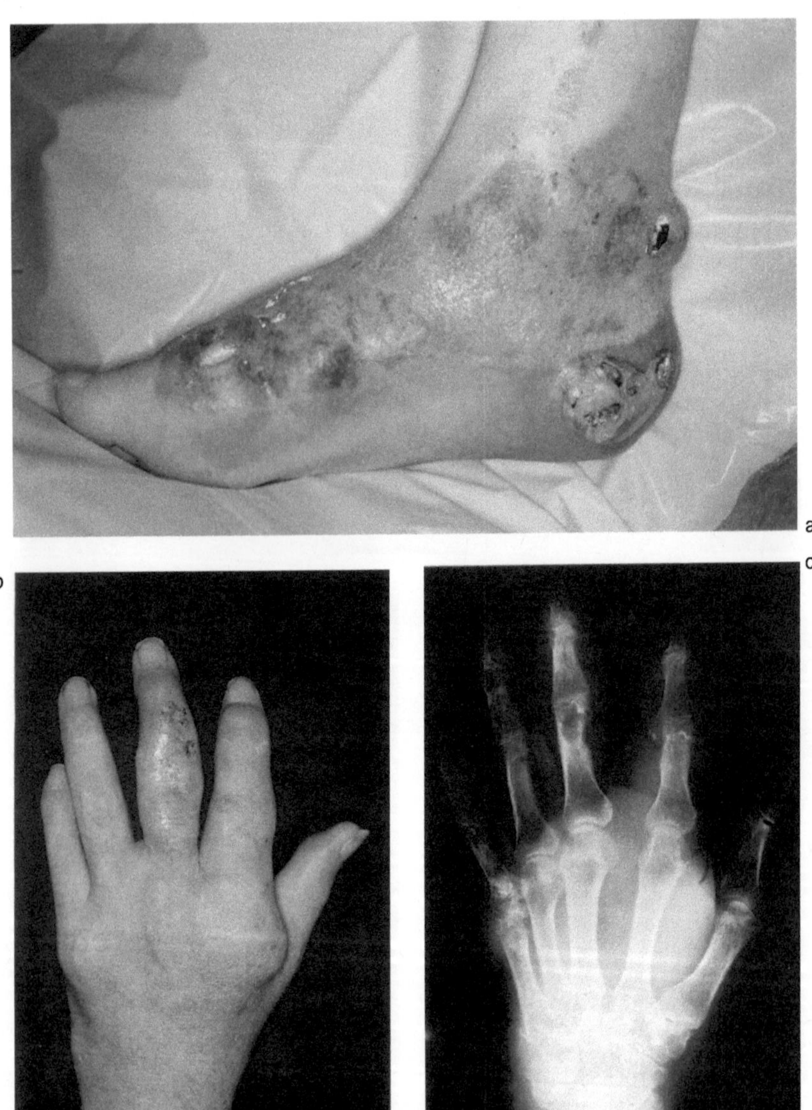

**Fig. 3a–c.** Right foot (**a**) and left hand (**b, c**) of a 48-year-old male with chronic gout caused by a defect in renal transport

**Table 2.** Hyperuricemia and gout: therapy aims and therapeutic possibilities

| Disease | Aims | Possibilities |
|---|---|---|
| **Acute gouty attack** | Freedom from pain<br>Avoidance of joint destruction | Therapy of pain with colchicine, nonsteroridal anti-inflammatory drugs or steroids<br>Physiotherapy |
| **Chronic hyperuricemia** | Avoidance of acute gouty attacks<br>Avoidance of tophi formation (joints, tendons, bursae, kidneys)<br>Maintenance of joint movability<br>Maintenance of working capacity and quality of life | Low purine diet<br>Physiotherapy<br>Drug therapy (Uricostatic, uricosuric or combined therapy)<br>Therapy of accompagning diseases (hyperlipidemia, hypertension, diabetes mellitus, obesity, alcoholism)<br>Normalisation of serum/plasma uric acid on levels between *5,0 – 6,0 mg/dl.* |

**Table 3.** Uric acid lowering drugs used in Europe [2, 3, 4, 7, 9]

| Group | Substance | Problems |
|---|---|---|
| Uricostatics | Allopurinol | Allergic and toxic reactions |
| Uricosurics | Benzbromarone | Genetic polymorphism in drug metabolism |
| | Sulfinpyrazone | Inhibition of platelet aggregation |
| | Probenecid | Inhibition of renal elimination of many drugs |
| | Irtemazole | Still in development |
| | (Fenofibrate | Primarily hypolipidemic) |

**Table 4.** Guidelines for the treatment with allopurinol or benzbromarone

**Xanthin oxidase inhibitor: allopurinol**

| | |
|---|---|
| Started at | 200–300 mg/day |
| Effective dose | 100–300 mg/day, rarely up to 600 mg/day |
| Renal insufficiency | Reduced excretion of oxipurinol, dose reduction necessary |

**Uricosuric drug: benzbromarone**

| | |
|---|---|
| Started at | 20 mg/day |
| Effective dose | 20–100 mg/day, rarely more |
| Renal insufficiency | Increasing loss of efficacy, below a creatinine clearance of about 25 ml/min ineffective |

**Table 5.** Definition of gout

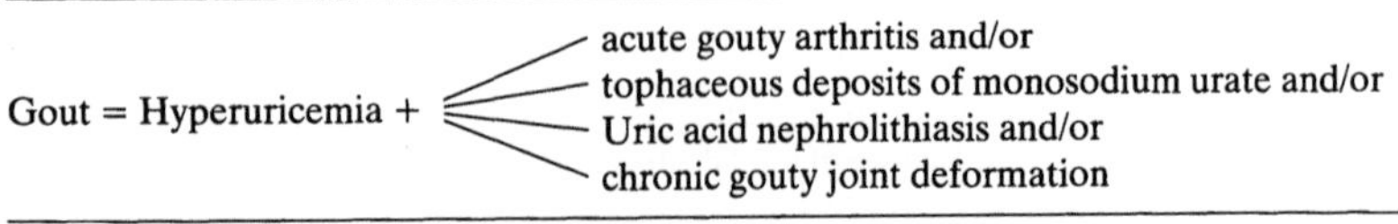


Patients suffer from chronic gout if they exhibit hyperuricemia and recurrent attacks of gouty arthritis leading to joint damage, tophaceous deposits of monosodium urate, nephrolithiasis, or kidney damage (Fig. 2, 3). Joint damage and tophi in bones can be diagnosed by X-ray or NMR [8].

Patients with hyperuricemia should be on a low purine diet. Uric acid lowering drug therapy is indicated if uric acid levels are constantly over 8.0 mg/dl, if patients suffer from gouty arthritis, urate nephrolithiasis or renal insufficiency, or if gouty arthritis or urate nephrolithiasis occur in the family of the patient. Uric acid lowering drugs can be either uricostatic (which reduce the formation of uric acid by inhibiting xanthin oxidase) or uricosuric (which increase renal uric acid excretion by inhibiting the tubular reabsorption of uric acid) (Tables 3 and 4), the drugs of choice being allopurinol and benzbromarone. In patients with severe hyperlipidemia, fenofibrate may be a helpful alternative.

# References


1. Gathof BS, Schreiber MA, Gresser U, Kamilli I, Zöllner N (1991) Importance of the confounding factors age and sex in the correlation of serum uric acid, cholesterol and triglyceride levels. Adv Exp Med Biol 309A: 231–234
2. Gresser U (1990) Therapie von Hyperurikämie und Gicht mit Urikosurika. Mediscript-Verlag, Munich
3. Gresser U (1993) Fenofibrate: a lipid lowering drug with uricosuric properties. Pharmaceutisch Weekblad (in press)
4. Gresser U, Adjan M, Zöllner N (1991) Deficient benzbromarone elimination from plasma: evidence for a new genetically determined polymorphism with an autosomal recessiv inheritance. Adv Exp Med Biol 309A: 157–160
5. Gresser U, Gathof BS (1991) Epidemiology of hyperuricaemia. In: Gresser U, Zöllner N (eds) Urate deposition in man and its clinical consequences. Springer, Berlin, Heidelberg, New York, pp 82–96
6. Gresser U, Gathof BS, Zöllner N (1990) Uric acid levels in Southern Germany 1989. A comparison with studies from 1962, 1971 and 1984. Klin Wochenschr 68: 1222–1228
7. Gresser U, Zöllner N (1989) Uricosuric effect of irtemazole in healthy subjects. Klin Wochenschr 67: 971–975
8. Kamilli I, Gresser U, Hahn D, Vogl T, Zöllner N (1992) Urate deposition: scintigraphy and magnetic resonance imaging. Br J Rheumatol 31: 135
9. Kamilli I, Gresser U, Pellkofer T, Löffler W, Zöllner N (1989) Uricosuric effect of irtemazole in hyperuricemic patients without and with renal insufficiency. Z Rheumatol 48: 307–312

# 2 The Biochemical Basis of Hyperuricemia and Gout

J. F. HENDERSON

Gout results from the precipitation of uric acid in joints. An understanding of the pathophysiology of gout begins with the fact that uric acid, a member of the purine family of metabolites, is only sparingly soluble in body fluids. In addition, normal concentrations of uric acid, at least in adult males, are not far above the point at which uric acid precipitates. The next point that needs to be understood is why uric acid concentrations become elevated to the point at which crystallization occurs. Factors that trigger precipitation also need to be understood, as body fluids sometimes remain supersaturated with uric acid for some time before crystallization begins. Finally, why uric acid crystallizes in joints needs to be understood (in the case of gout), as it can and does also crystallize out in other tissues as well. Factors that result in elevated uric acid concentrations will be considered here.

## Potential Causes of Hyperuricemia

Simple logic leads to the listing of five potential causes of hyperuricemia that may, in turn, lead to gout.

1. Increased consumption of purine-rich foods; virtually all of the purines contained in food are converted to uric acid and excreted.
2. Decreased excretion of uric acid by the kidneys, a complex process in which many things can go wrong. Clinically, this may be the most important cause of hyperuricemia.
3. Accelerated death or turnover of cells. Upon the death of cells their nucleic acid and soluble purines will be converted to uric acid. This is seen, for example, following cancer chemotherapy or radiotherapy, or in psoriasis or hemolytic anemia.
4. Biochemical changes within cells may lead to accelerated breakdown of purines, leading to increased production of uric acid.
5. Other biochemical changes may lead to accelerated rates of purine biosynthesis, also leading to increased production of uric acid.

Clearly the last two types of causes of hyperuricemia are of the greatest interest to biochemists.

# Increased Purine Breakdown

One well-recognized cause of increased purine breakdown is hereditary fructose intolerance, which is due to the absence of fructose-1-phosphate aldolase. When such patients eat fructose-containing foods, the fructose is phosphorylated by adenosine triphosphate (ATP), but the resulting fructose-1-phosphate cannot be metabolized to support the resynthesis of ATP by the glycolytic pathway. At least part of the adenosine diphosphate (ADP) so formed then breaks down completely to uric acid.

Any condition, in fact, that leads to impaired energy generation may lead to some ATP breakdown and hence to increased uric acid production. These may include decreased lung function, decreased availability of phosphate, or any condition in which there is hypoxia, ischemia, or shortage of metabolizable substrates. These conditions may or may not be servere enough to lead to clinical gout, however.

This area continues to challenge biochemists. The regulation of nucleotide dephosphorylation in intact cells and tissues, especially the dephosphorylation of nucleoside monophosphates, is not clearly understood. In addition, there are major differences among tissues in the relative rates of alternative pathways of ATP breakdown. Finally, guanosine triphosphate (GTP) breakdown also has to be considered.

# Accelerated Purine Biosynthesis

The biosynthesis of purines is a complex process. Its rate limiting step involves glutamine and the sugar derivative, phosphoribosyl pyrophosphate or PPRP. As glutamine usually is in plentiful supply, the concentration and availability of PPRP are critical determinants of the rate of purine biosynthesis. Studies over the last two decades have identified several conditions in which gout is due to elevated PPRP.

When the pathway of purine biosynthesis is accelerated, the first purine nucleotide that is formed is inosinic acid or inosinate. This may be metabolized along three separate pathways, two of which are conversion to adenine and guanine nucleoteides. When inosinic acid synthesis is accelerated, however, little additional conversion to adenine and guanine nucleotides occurs. Instead, most of the excess is dephosphorylated and converted by a series of reactions to uric acid.

# Increased PPRP

Increased availability of PPRP may result from two types of biochemical changes. First, decreased utilization by other pathways that use PPRP. Second, by increased biosynthesis of PPRP.

Decreased utilization of PPRP by alternative pathways needs to be considered because this sugar phosphate is involved in so many reactions. It donates ribose phosphate not only to glutamine, but also to adenine, hypoxanthine, and guanine in the area of purine metabolism. In pyrimidine metabolism it reacts with orotic acid and with uracil, and in pyridine metabolism it reacts with nicotinamide and nicotine acid.

Decreased utilization may result, for example, from complete or partial loss of hypoxanthine guanine phosphoribosyltransferase (HGPRT). In addition, certain patients have HGPRT molecules that exhibit decreased affinity for PPRP and hence show decreased total effective enzyme at normal concentrations of this substrate.

Failure to utilize PPRP in the HGPRT reaction, to name the most well-studied examples, allows this sugar phosphate to be used by other reactions, including purine biosynthesis de novo.

Increased biosynthesis of PPRP has been shown in some patients to result from increased amounts of the responsible enzyme, PPRP synthetase. In others, the enzyme has increased affinity for inorganic phosphate, which is a required allosteric activator of this enzyme.

PRPP availability may also be increased if ribose-5-phosphate synthesis is accelerated along either the oxidative or nonoxidative pentose phosphate pathways. In glucose-6-phosphate deficiency, the inability to metabolize glucose-5-phosphate by this reaction leads to increased utilization along all other possible pathways: to glycogen, to lactate, and to ribose-5-phosphate. This in turn can lead to increased PPRP and to accelerated uric acid biosynthesis.

## Decreased Feedback Regulation

Another type of biochemical change that has been postulated as a cause of hyperuricemia and demonstrated in some experimental systems, is decreased feedback regulation of purine synthesis. This might be caused, for example, by lowered concentrations of ATP and GTP or other effectors.

Study of this possibility is complicated by the fact that these endproduct nucleotides inhibit two consecutive steps in the process: PRPP synthetase itself, and then the first reaction of purine biosynthesis, amidophosphoribosyltransferase.

## Multiple Causes of Hyperuricemia

Other biochemical changes may lead to hyperuricemia in more than one way. For example, a partial deficiency of inosinic acid dehydrogenase may be expected to lead to increased breakdown of this substrate. In addition, such an enzyme deficiency may be expected to result in lowered concentrations of GTP. This in turn may give rise to accelerated purine biosynthesis due to decreased feedback inhibition, as just discussed.

## Future Prospects

The complexity of purine metabolism in the human body leads us to expect the future discovery of additional biochemical abnormalities leading to hyperuricemia. These may be difficult to elucidate and study, however, if they involve: (a) partial deficiencies rather than total loss of a process, (b) tissues that are not easily studied experimentaly, and (c) changes in the supply of substrates for energy metabolism.

# 3 The Genetic Basic of Hyperuricemia and Gout

B. J. ROESSLER

Gout and an inherited predisposition to the disease have been recognized for centuries. Since 1950 treatments have been developed to modify the natural history of the disease, the biochemical basis of purine biosynthetic systems have been elucidated, and in the past several years, some causes of gout have been defined at the molecular level. Despite these accomplishments, the metabolic and genetic basis for gout in the majority of patients remains unclear.

Evidence for the genetic control of serum urate levels exists on several levels. Although there is considerable variability of plasma urate values among various population samples, certain epidemiological information suggests that genetic factors are important in the determination of serum urate levels [1]. For example, indigenous Pacific islanders have a significantly higher mean serum urate level than that found in other ethnic populations [2, 3]. In many of these population studies, hyperuricemia appears to be multifactorial and attributable to a combination of genetic and nongenetic factors. The genetic factors appear to be polygenic [4]. Studies in Blackfoot and Pima Indians indicate that some of the genes involved in polygenic transmission function as autosomal dominants while others function as X-linked dominants [2]. Undoubtedly the genetic factors responsible for determining hyperuricemia in any individual include both cumulative gene action and single gene effects.

Primary gout is attributable to inborn errors of either purine biosynthetic metabolism or specific defects in the renal excretion of uric acid. This category is both biochemically and genetically heterogenous. The vast majority (up to 90%) of gouty patients represent those in whom the fractional renal urate clearance is reduced. In many of these patients this defect is the major or even sole basis of hyperuricemia, although the molecular events responsible for these changes have not been defined.

Similar defects resulting in the underexcretion of uric acid may be present in certain enzymopathic overproducers of uric acid, for example, in those with glucose-6-phosphatase (G6P) deficiency or other glycogenoses [2, 5]. Defects in renal urate excretion have not been identified in patients with hypoxanthine guanine phosphoribosyltransferase (HGPRT) deficiency or phosphoribosyl-pyrophosphate synthetase (PRPS) superactive variants.

Parenchymal renal diseases that cause alterations in the glomerular or tubular handling of uric acid, may result in the development of secondary hyperuricemia and, rarely, gouty arthritis. In most cases this is thought to be a result of nonspecific tubular damage not directly related to genetic abnormalities in the renal handling of uric acid.

Amongst those with primary metabolic derangements, the largest subgroup consists of patients in whom the biochemical defect is as yet undefined, but pedigree analysis suggests an inherited cause. The remainder include those with partial or complete HGPRT deficiency or PRPS superactivity [2, 6, 7]. Secondary metabolic gout is attributable to inborn errors of metabolism, specifically G6P deficiency states.

The metabolic basis for most of the abnormalities in the renal handling of uric acid remains to be determined, although kindred analysis continues to suggest a genetic etiology. For example, studies by Scott and Pollard described a graded correlation between uric acid clearance values in patients with gout and the mean values determined in their first degree relatives [8]. This correlation was closer in male relatives than in female relatives. Additionally a recent study by Emmerson et al. compared the renal clearance of urate in groups of both monozygotic and dizygotic twins [9]. This study found that correlation for monozygotic twins was 0.85 while that for dizygotic twins was 0.54, a statistically significant difference with $p = 0.04$. The authors performed a heritability analysis on these data which was estimated to be 60% with a confidence interval of 40%–100%. The data lend support to the hypothesis that genetic factors strongly influence the renal clearance of urate and in turn determine the serum urate concentration.

Since the molecular basis of urate transport in humans remains ill defined, definition of the molecular genetics responsible for primary renal hyperuricemia remain largely unknown. As the details of renal urate handling are elucidated and the specific transport mechanisms are identified the ability to determine the molecular basis of hypouricosuria will progress rapidly.

Additionally, insight into the possible genetic mechanisms of abnormal renal clearance of urate may be gained by the continued study of several inherited diseases of the kidney that phenotypically express hyperuricemia and gout. Table 1 outlines those inherited renal disease and identifies what is known regarding the mode of inheritance. These diseases include; adult polycystic kidney disease, medullary cystic disease, focal tubulointerstitial disease, and familial urate nephropathy [10–16].

**Table 1.** Renal disease associated with hyperuricemia and gout

| Disease | Inheritance | Incidence | Mechanism |
|---|---|---|---|
| Adult polycystic | Autosomal dominant (chromosome 16) | Common | Ion pump(?) |
| Adult medullary cystic | Autosomal dominant | Rare | Tubular damage |
| Focal tubulointerstitial | Autosomal dominant | Rare | Tubular damage |
| Familial urate nephropathy | Autosomal dominant | Rare | Urate transport(?) |

Up to 30% of patients with adult polycystic kidney disease develop gout. Hypouricosuria is often an early manifestation of the renal lesion, and hyperuricemia and gout precede the development of renal failure [10]. Although the basis for this association is unclear, altered renal handling of uric acid may be an early manifestation of the underlying molecular defect. The identification and cloning of the gene associated with polycystic kidney disease will allow this hypothesis to be systematically evaluated.

Adult onset medullary cystic kidney disease is an autosomal dominant condition characterized by the development of cysts in the medulla and corticomedullary junction [11]. The progressive tubular atrophy is thought to be responsible for the reduction in renal urate clearance, and there is little evidence to suggest a primary defect in uric acid transport, although cystic kidney disease are thought to be related to abnormalities in cellular ion pumps.

Focal tubulointerstitial disease is a rare autosomal dominant condition initially reported by Leumann and Wegmann in 1983 (12). The authors studied a single family and believe that the distinct histologic findings, including focal tubulointerstitial lesions, glomerular sclerosis, giant mitochondria, and focal interruptions of the tubular basement membrane, allow them to distinguish this condition from those kindred with familial urate nephropathy. Given the heterogeneity of phenotypic expression that can result from different mutations occurring in a single gene, it is unlikely that this disease can be positively distinguished from familial urate nephropathy without the use of genetic analysis.

Familial urate nephropathy most likely represents a disease or closely related group of diseases in which the primary genetic defect may be the result of an abnormality in renal urate transport [13–16]. This condition is probably heterogenous with regard to either the gene involved and/or the type of genetic mutation present. Over 20 affected kindred have been reported [17]. Studies indicate that the low fractional excretion of urate (3%–4% of normal) precedes the other manifestations of the disease. Recent studies by a number of investigators have hypothesized several mechanisms of abnormal urate transport of explain the clinicopathological findings in this condition. Further study of renal tissue from these kindred may allow the identification and isolation of urate transport proteins in the brush border or basolateral membrane that would be targets for cDNA cloning and genetic sequence analysis in affected individuals.

Alternatively, direct genetic analysis of these kindred allow us to gain further insight into the genes responsible for determining urate transport and the fractional excretion of uric acid. The application of the methods of linkage analysis and positional cloning to the affected kindred may be means of identifying the gene involved in determining the abnormal phenotype. Similar methods of genetic analysis have already been successfully employed to identify the genes responsible for chronic granulomatous disease, Duchenne muscular dystrophy, retinoblastoma, cystic fibrosis, and neurofibromatosis [18–20].

As an example, these methods are currently being employed to identify the gene responsible for adult polycystic kidney disease. Cosmid walking and chromosome jumping have been used to identify gene loci closely linked to the gene for polycystic kidney disease and have localized it to the region p13.3 of chromosome 16 [21, 22]. Several candidate genes have now been isolated and are undergoing additional genetic analysis. Interestingly, a recently identified candidate gene has homology to the proteolipid component of a proton channel of the vaculor proton-ATPase [22. It is interesting to consider how these ion pumps may be related to tubular mechanisms of uric acid transport.

Linkage analysis has previously been established in kindred exhibition both familial hyperuricemia and familial hypertriglyceridemia [2]. Similar linkage has been noted in some kindred with familial nephropathies [2]. If suitably linked marker genes and chromosomal probes can be identified, then these methods would be potentially applicable to the study of kindred affected by familial juvenile gouty nephropathy.

Much more is known regarding the genetics of purine biosynthetic enzymopathies. To date, three distinct genetically determined enzymatic defects have been defined that lead to marked hyperuricemia and gout, although it is probable that other defects of purine metabolism resulting in hyperuricemia will be identified by genetic analysis. Those currently identified are partial and complete deficiency of HGPRT, superactive variants of PRPS and deficiency states of GGP [2]. Table 2 outlines these disease states, their modes of inheritance, and mechanism for producing hyperuricemia. HGPRT deficiency and PRPS superactivity have both been characterized at the genetic level.

**Table 2.** Enzymopathies associated with hyperuricemia and gout

| Disease | Inheritance | Incidence | Mechanism |
| --- | --- | --- | --- |
| HGPRT (partial and complete) | X-linked | 1:100 000 | Overproduction |
| PRS superactive variants | X-linked | Rare | Overproduction |
| G6P (Ia) | Autosomal recessive | 1:1–400 000 | Overproduction Underexcretion |
| G6P(Ib) | Autosomal recessive | Rare | Overproduction Underexcretion |

HGPRT, hypoxanthine guanine phosphoribosyltransferase; PRS, phosphoribosylpyrophosphate synthetase, G6P, glucose-6-phosphatase.

In the case of G6P deficiency states (glycogen storage disease type Ia) the gene for the enzyme has yet to be identified and cloned. Neither has the defect in the microsomal transport system responsible for the disease in some individuals been identified at the genetic level (type Ib). In the case of G6P deficiency states, accelerated adenosine triphosphate (ATP) turnover has been clearly shown to be the basis for overproduction of urate in this disorder [2].

A similar mechanisms of accelerated ATP turnover produces the exertional hyperuricemia observed in several genetic conditions in which muscle metabolism is impaired. These include glycogen storage disease types III, V and VII [2]. The gene and cDNA encoding for myophosphofructokinase have been cloned [23, 24]. Recently this sequence information has been used to identify a mutation responsible for a case of type VII glycogen storage disease. Nakajima et al. identified a point mutation occurring at the splice donor site of intron 15 of the muscle phosphofructokinase gene that resulted in the deletion of 75 base pairs (bp) in the processed mRNA [25]. This in turn results in the deletion of 25 amino acid residues from the muscle phosphofructokinase enzyme rendering it catalytically inactive. Similar genetic analyses of the other muscle glycogenoses will undoubtedly lead to an improved understanding of the molecular events that lead to myogenic hyperuricemia.

Since the glycogenoses share a biochemical mechanism responsible for the development of hyperuricemia and because subtle defects in the renal handling of urate in these disease states have been identified [26, 27], the further elucidation of the molecular basis of these defects may produce additional clues that will lead to an improved understanding of the renal metabolism of uric acid.

HGPRT deficiency states have been some of the most extensively studied genetic diseases [28–31]. In addition to providing protein structure-function information, the genetic analysis of HGPRT deficiency states can also provide us with insight into general mechanisms of human biology. Genetic studies have revealed that not all HGPRT deficiency states are caused by point mutations with resultant structural enzyme abnormalities. Included are deletions, duplications, and insertions that result in improperly processed HGPRT mRNA or HGPRT mRNA that cannot be translated.

Superactive variants of PRPS result in the phenotypic expression of hyperuricemia and gout by increasing the intracellular amount of phosphoribosylpyrophosphate (PRPP) that drives the de novo production of purines [2, 7]. Examination of the catalytic properties of partially purified PRPS from affected patients reveals that the abnormalities are functionally heterogenous [7]. The genetic regulation of normal and mutant PRPS expression appears to be very complex and involves multiple PRPS loci. It follows that the genetic mutations responsible for these variants are complex and heterogenous but our preliminary experience suggests that they may fall into identifiable subgroups.

Two of the genes that encode for PRPS exist as distinct loci located on separate arms of the X chromosome, and have been designated PRPS1 and PRPS2 [32, 33]. A second autosomal locus has been identified on chromosome 9 and designated PRPS3. Each locus is transcriptionally active, as indicated by northern blot analysis of human tissues, although there is evidence for tissue specificity in the expression of the PRPS isoform mRNAs [32, 34]. The proteins produced by the PRPS1 and 2 genes are extensively homologous but do differ by 14 amino acids scattered throughout their length. Expression of each of these human isoforms in a bacterial expression system has shown preliminary results indicating each isoform is functionally active [35].

In order to identify the mutations that are responsible for the phenotypic expression of PRPS superactivity we have performed sequence analysis of both

PRPS1 and PRPS2 mRNAs using the method of direct sequencing of amplified transcripts on the cell lines obtained from affected patients. To date, two distinct point mutations occurring in the coding region of the PRPS1 gene associated with the phenotypic expression of PRPS superactivity [36]. In each case the phenotype is characterized by a PRPS enzyme that is resistant to feedback inhibition by purine nucleotides.

In one patient an A to G transition at nucleotide 341 has been identified. In another patient a C to G transition at nucleotide 547 has been identified. This predicts an asparagine to serine change at amino acid 113 in the first patient and an aspartic acid to histidine change at amino acid 182 in the second patient.

We are currently in the process of completing cloning of these mutant PRPS1 cDNAs into bacterial expression systems designed to efficiently express normal human PRPS1 and PRPS2 isoforms. Analysis of the recombinant mutant proteins will allow us to establish a more precise relationship between these point mutations in the PRPS1 gene and the distinctive purine nucleotide feedback resistant phenotypes found in the affected individuals.

In conclusion, the biochemical and genetic basis of hyperuricemia and gout has been defined in a small percentage of patients, specifically those with rare purine enzymopathies including HGPRT deficiency states and PRPS superactive variants. In the majority of patients, hyperuricemia and gout is a result of hypouricosuria. A complete understanding of the genetic aspects of hyperuricemia in these kindred awaits the definition of the exact biochemical mechanisms responsible for the renal metabolism of uric acid. The techniques of molecular genetics will continue to improve our ability to elucidate these abnormalities.

# References


1. Stecher RM, Hersh AH, Soloman WM (1949) The heredity of gout and its relationship to familial hyperuricemia. Ann Int Med 31: 595–600
2. Palella TD, Fox IH (1989) Hyperuricemia and gout. In: Stanbury JB; Wyngaarden JB; Fredrickson DS, Goldstein JL, Brown MS (eds). The metabolic basis of inherited disease, 6th ed. McGraw Hill, New York, chap 37
3. Harlan WR, Hull AL, Schmouder RP, Thompson FE, Larkin FA, Landis JR (1983) Dietary intake and cardiovascular risk factor, Part II. Serum urate, serum cholesterol, and correlates. In: Vital and Health Statistics, United States 1971–1975. Series II, No. 227. DHHS Pub. No. (PHS) 83–1677. Public Health Service, Washington, D.C.
4. Neel JV, Rakic MT, Davidson RI, Valkenburg HA, Mikkelsen WM (1965) A reconsideration of the distribution of serum uric acid values in the families of Smyth, Cotterman, and Freyberg. Am J Hum Genet 17: 14–20
5. Kelley WN, Rosenbloom FM, Seegmiller JE, Howell RR (1968) Excessive production of uric acid in type I glycogen storage disease. J Pediatr 72: 488–496
6. Kelley WN, Rosenbloom FM, Henderson JF, Seegmiller JE (1969) Hypoxanthine-guanine phosphoribosyltransferase deficiency and gout. Ann Intern Med 70: 155–206
7. Losman M, Rimon D, Kim M, Becker MA (1985) Selective expression of phosphoribosylpyrophosphate synthetase superactivity in human lymphoblast lines. J Clin Invest 76: 1657–1664

8. Scott JT, Pollard AC (1970) Uric acid excretion in the relatives of patients with gout. Ann Rheum Dis 29: 397–400

9. Emmerson BT, Nagel SL, O'Connor J, Duffy DL, Martin NG (1991) The genetics of renal excretion of urate in man. Adv Exp Biol Med (in Press)

10. Rivera JV (1974) Gout and polycystic kidney disease. Ann Intern Med 80: 427–435

11. Thompson GR, Weiss JJ, Goldman RT (1978) Familial occurrence of hyperuricemia, gout and medullary cystic disease. Arch Intern med 138: 1614–17

12. Leumann EP, Wegmann W (1983) Familial nephropathy with hyperuricemia and gout. Nephron 34: 51–55

13. Simmonds HA, Cameron JS, Potter CF, Warren D, Gibson T, Farebrother D (1980) Renal failure in young subjects with familial gout. Adv Exp Med Biol 122A: 15–20

14. Simmonds HA, Warren DJ, Cameron JS (1980) Familial gout and renal failure in young women. Clin Nephrol 14: 176–181

15. Stapleton FB, Nyhan WL, Borden M (1981) Renal pathogenesis of familial hyperuricemia: studies in two kindreds. Pediatr Res 15: 1447–53

16. MacDermot KD, Allsop J, Watts RW (1984) The rate of purine synthesis de novo in blood mononuclear cells in vitro from patients with familial hyperuricemic nephropathy. Clin Sci 67: 249–58

17. Cameron JS, Moro F, Simmonds HA (1991) What is the pathogenesis of familial gouty nephropathy? Adv Exp Biol Med 309A: 185–9

18. Collins FS, Riordan JR, Tsui LC (1990) The cystic fibrosis gene: isolation and significance. Hosp Pract 25: 48–57

19. Rommens JM, Iannuzzi MC, Kerem B, Drumm ML, Melmer G, Dean M, Rozmahel R, Cole JL, Kennedy D, Hidaka N, et al. (1989) Identification of the cystic fibrosis gene: chromosome walking and jumping. Science 245: 1059–65

20. Fountain JW, Wallace MR, Bruce MA, Seizinger BR, Menon AG, Gusella JF, Michels VV, Schmidt MA, Dewald GW, Collins FS (1989) Physical mapping of a translocation breakpoint in neurofibromatosis. Science 244: 1085–7

21. Gillespie GAJ, Germino GG, Somlo S, Weinstat-Saslow D, Breuning MH, Reeders ST (1990) Cosmid walking and chromosome jumping in the region of PKD1 reveal a locus duplication and three CpG islands. Nucleic Acid Res 18(23): 7071–7075

22. Gillespie GAJ, Somlo S, Germino GG, Weinstat-Saslow D, Reeders ST (1991) CpG island in the region of an autosomal dominant polycystic kidney disease locus defines the 5'end of a gene encoding a putative proton channel. Proc Natl Acad Sci USA 88: 4289–4293

23. Nakajima H, Noguchi T, Yamasaki T, Kono N, Tanaka T, Tarui S (1987) Cloning of human muscle phosphofructokinase cDNA. FEBS Lett 223: 113–6

24. Valdez BC, Chen Z, Sosa MG, Younathan ES, Chang SH (1989) Human 6-phosphofructo-1-kinase gene has an additional intron upstream of start codon. Gene 76: 167–9

25. Nakajima H, Kono N, Yamasaki T, Hotta Y, Kawachi M, Kuwajima M, Noguchi T, Tanaka T, Tarui S (1990) Genetic defect in muscle phosphofructokinase deficiency. J Biol Chem 265: 9392–9395

26. Howell PR (1965) The interrelationship of glycogen storage disease and gout. Arth Rheum 8: 780–787

27. Howell RR, Ashton DM, Wyngaarden JB (1962) Glucose 6 phosphatase deficiency glycogen storage disease. Studies on the interrelationship of carbohydrate, lipid and purine abnormalities. Pediatrics 29: 553

28. Wilson JM, Stout JT, Palella TD, Davidson BL, Kelley WN, Caskey CT (1986) A molecular survey of hypoxanthine-guanine phosphoribosyltransferase deficiency in man. J Clin Invest 77: 188–192

29. Davidson BL, Tarle SA, Palella TD, Kelley WN (1989) The molecular basis of HPRT deficiency in ten subjects determined by direct sequencing of amplified transcripts. J Clin Invest 84: 342–346
30. Davidson BL, Tarle SA, van Antwerp ME, Gibbs DA, Watts RE, Kelley WN, Palella TD (1991) Identification of seventeen independent mutations responsible for human HPRT deficieny. Am J Hum Genet 48(5): 951–958
31. Tarle SA, Davidson BL, Wu VC, Zidar FJ, Seegmiller JE, Kelley WN, Palella TD (1991) Determination of the mutations responsible for the Lesch-Nyhan syndrome in seventeen subjects. Genomics 10: 499–501
32. Roessler BJ, Bell G, Heidler S, Seino S, Becker MA, Palella TD (1989) Cloning of two distinct copies of human phosphoribosylpyrophosphate synthetase cDNA. Nucleic Acids Res 18: 193
33. Becker MA, Heidler SA, Bell GI, Seino S, LeBeau MM, Westbrook CA, Newman W, Shapiro LJ, Roessler BJ, Palella TD (1990) Cloning of cDNAs for human PRPP synthetastes 1 and 2 and regional localization of the PPRS1 and PPRS2 genes on the X chromosome. Genomic 8: 555–561
34. Taira M, Iizasa T, Yamada K, Shimada H, Tatibana M (1989) Tissue differential expression of two distinct genes for phosphoribosylpyrophosphate synthetase and existence of the tests specific transcript. Biochem Biophys Acta 1007: 203–208
35. Becker MA, Heidler SA, Nosal JM, Switzer RL, LeBeau MM, Shapiro LJ, Palella TD, Roessler BJ (1991) Human phosphoribosylpyrophosphate synthetase (PRS) 2: an independent active, X-chromosome-linked PRS isoform. Adv Exp Med Biol 309B: 129–32
36. Roessler BJ, Golovoy N, Palella TD, Heidler S, Becker MA (1991) Identification of distinct PRS1 mutations in two patients with X-linked phosphoribosylpyrophosphate synthetase superactivity. Adv Exp Med Biol 309B: 125–128

# III Immunodeficiency Disease: Adenosine Deaminase (ADA) and Purine-Nucleoside Phosphorylase (PNP) Deficiencies

# 1 Introductory Remarks

B. S. MITCHELL

The discovery in the 1970s by Dr. Eloise Giblett of two inborn errors of purine metabolism which are etiologically associated with immunodeficiency disease constituted a major step forward in understanding the basis of lymphocyte dysfunction in these individuals. The association of adenosine deaminase deficiency with both T and B cell dysfunction has spawned a large literature on the metabolic consequences of this enzyme deficiency state as well as extensive investigations into the expression of the adenosine deaminase gene and the structural alterations in adenosine deaminase which give rise to the deficiency state. Similarly, the association of purine-nucleoside phosphorylase deficiency with an isolated T cell defect has led both to the delineation of a number of mutations within this protein which are associated with the deficiency state and to an extensive search for potent and selective inhibitors of this enzyme which might be useful as immunosuppressive agents. A major consequence of these investigations has been the development of effective new therapies, including bone marrow transplantation and the use of polyethylene glycol (PEG)-conjugated adenosine deaminase, for treating these disorders. In the following chapters, the clinical aspects of adenosine deaminase deficiency and the results of bone marrow transplantation for this disorder in patients treated at the Universitäts-Kinderklinik in Ulm are discussed by Dr. Friedrich. Professor Gutensohn will review the pathogenesis of the immunodeficiency in these syndromes. Finally, Dr. Hershfield will present his data on adenosine deaminase deficient children in relation to what it know of the structure of the protein. He will also discuss the clinical and biochemical results of PEG-adenosine deaminase administration to these children.

# 2  The Clinical Aspects of ADA and PNP Deficiencies

W. Friedrich and W. Hartmann

## Introduction

Severe combined immunodeficiency disease (SCID) constitutes the most profound form of a congenital abnormality of the immune system. Infants with this disorder lack both T cell and B cell functions and development of life threatening infections early in life is a prominent clinical finding (Rosen et al. 1984). The disorder usually is secondary to a global defect of lymphoid differentiation and more rarely is caused by a functional abnormality of more mature effector cells. As based on immunological, clinical, genetic, and enzymatic findings, several variants of SCID can be differentiated. The best characterized variants are those associated with abnormalities of purine metabolism. Giblett and coworkers (Giblett et al. 1972) described two female infants who presented with autosomal recessive SCID and who also laked the enzyme adenosine deaminase (ADA). The same workers also described the first patient with purine nucleoside phosphorylase (PNP) deficiency and immunological abnormalities (Giblett et al. 1975). Clear evidence was presented for a causal relationship between the immunodeficiency and the enzymatic defect, leading to the accumulation of toxic metabolites that are inhibitory in particular for lymphoid cells (Cohen et al. 1978; Hirschhorn et al. 1979)

In this presentation, an overview of the clinical and pathological findings in patients with ADA and with PNP deficiency will be given as well as an outline of currently available curative treatment by bone marrow transplantation (BMT). This will be based primarily on our own experience in several affected patients.

## Incidence and Clinical Presentation of Congenital Immunodeficiency Associated with Enzymatic Abnormalities

### SCID with ADA Deficiency

Of a total group of 58 patients with SCID admitted to our institution since 1981, five were found to have ADA deficiency as based on finding extremely low or absent enzyme activities in red blood cells. These activities were normal in the other 53 patients, resulting in an overall prevalence of about 10% in our patient population. In the literature, a prevalence of about 20% has been reported, as based on a survey of 130 patients with SCID (Hirschhorn 1979).

Prominent anamnestic and clinical findings in patients with ADA deficiency as also observed in our patients are:

- Complications from infections early in life in particular bronchopneumonia, recurrent and chronic gastroenteritis with failure to thrive, septicemia fungal disease
- Hypoplastic lymphoid tissue, no tonsils, no palpable lymph nodes, and radiologically absent thymic tissue
- Skeletal abnormalities evident as a palpable prominence of the costochondral rib junction and radiologically as flaring and cupping of the costochondral junctions, a dysplastic pelvis, and abnormal vertebral bodies

Our first patient, who was born as a healthy infant at term to unrelated parents, developed at the age of 3 weeks enteritis, vomiting, and a cough. Some 2 weeks later, his clinical status deterioated dramatically and he was diagnosed as suffering from bacterial sepsis. With antibiotic treatment, the infection was controlled, but symptoms of gasteroenteritis persisted and the child lost a significant amount of weight. Rotavirus was recovered from stool cultures. On day 14 of his hospital course, acute pneumonia caused by *Pneumocystic carinii* developed, which was successfully treated by trimethoprim and sulfisoxazole. These unusual infectious complications prompted an immunological evaluation, which revealed profound deficiencies of lymphocyte numbers and functions, consistent with SCID. In addition to an abnormally narrow upper mediastinum and a translucent substernal space noted in chest X-rays, findings consistent with a small or absent thymus, the radiologist also pointed out distinct skeletal abnormalities, in particular of the ribs, with abnormal cupping of costochondral junctions suggestive of ADA deficiency (Wolfson and Cross 1975; Cederbaum et al. 1976). This diagnosis was confirmed by red cell enzymatic studies. For further treatment the child was referred to us at the age of 11 weeks and was in fair clinical condition. Neurologically, he was noted to be mildly hypotonic with poor head and trunk control.

Anamnestic histories in three other patients with ADA deficiency were fairly similar, with early onset of nonremitting gastroenteritis secondary to viral infections, symptoms of bronchopulmonary infections, and a rapidly detoriating state of well-being during the first months of life. In one patient, the sister of our first patient the diagnosis of ADA deficiency had been established prenatally. The parents, while initially determined to interrupt the pregnancy in case of positive findings, later on had changed their minds. In the newborn the diagnosis was confirmed and she was kept in a clean environment in the hospital until transfer to our hospital at the age of 7 weeks. In spite of these precautions she had developed pulmonary and intestinal infections. Radiological studies, in contrast to those of the brother, did not reveal gross skeletal abnormalities.

## PNP Deficiency

Immunodeficiency in association with PNP deficiency was found in one patient. The child had suffered from early childhood on from recurrent and chronic infections of the upper respiratory tract, involving the middle ears, sinus and bronchial system, but had required no hospitalizations. Psychomotor development was mildly retarded without obvious etiology. At 4 years of age he had serologically proven Epstein-Barr virus (EBV) monucleosis, which took an uneventful course. IgG serum levels at that time were slightly below the lower age limit, while IgA and IgM were normal. At 5 years of age he developed severe, near fatal complications from varicella with formation of large hemorrhagic skin blisters, encephalitis, iridocyclitis, and a very protracted course in spite of systemic treatment with acyclovir over several weeks and the use of interferon-β. Immunological evaluation then showed normal immunoglobulin serum levels, but low total lymphocyte counts (below 1000/μl) with a markedly reduced percentage of T cells and an almost absent proliferative response upon mitogen stimulation in vitro. PNP enzyme activity in red cells was absent. The child was presented to us several weeks later for possible treatment by BMT. Unfortunately, while awaiting confirmation of a transplant from an unrelated marrow donor, the child developed a metastatic Burkitt's lymphoma and died without undergoing further treatment.

# Laboratory Findings in Patients with ADA Deficiency

Table 1 shows the results of immunological and biochemical studies in the five patients with ADA deficiency. Enzyme activities were extremely low, while in the parents these were usually intermediate, consistent with heterozygous carrier states. Of note, severe lymphocytopenia involving both T cells and B cells was present in all five patients, in particular in patient 4, the sister of patient 1, who was examined shortly after birth. None of the patients had any evidence of humoral immune functions, as reflected by extremely low IgA and IgM serum levels as well as low IgG, which continued to decline prior to substitution. Also there was no evidence of residual cellular immunity, including in the child studied shortly after birth.

We also examined three patients in regard to the possibility of an engraftment by maternal lymphocytes secondary to intrauterine maternofetal transfusions. These studies were prompted by observations in other SCID patients, in whom we observed this complication in more than 50% even when circulating T cells were extremely low. There was no such evidence in the blood and the bone marrow in these patients.

Upon radiological and ultrasound examinations, no thymic tissue was identified in any of the patients. Skeletal abnormalities of the ribs, as noted above, were present in three additional patients. We have no observed similar skeletal findings in any of our other SCID patients with normal ADA. In three patients histo-

**Table 1.** Laboratory findings in five patients with ADA deficiency

|  | Patient 1 | Patient 2 | Patient 3 | Patient 4 | Patient 5 |
|---|---|---|---|---|---|
| Lymph/µl | 500 | 450 | 600 | 300 | 200 |
| CD3 positive (T cells) % | 0 | 0 | 27 | 10 | 2 |
| CD20 positive (B cells) % | 0 | 0 | 15 | 0 | 0 |
| In vitro lymphocyte response | Absent | Absent | Absent | Absent | Absent |
| IgG (g/l) | 1.0 | 1.15 | 0.9 | 4.0 | 1.4 |
| IgM (g/l) | 0 | 0 | 0 | 0 | 0 |
| IgA (g/l) | 0 | 0 | 0 | 0 | 0 |
| ADA activity in red cells (U/I EC)[a] | 0.15 | 0.03 | 0.02 | 0.04 | 2.0 |
| Skeletal abnormalities | Yes | Yes | No | Yes | Yes |

[a] Normal values: 410–630 U/E EC

morphological studies were performed on bone biopsies obtained from iliac crests. In all three, striking histomorphological abnormalities were noted with grossly abnormal structures of growth zones and pronounced vacuolization of chondrocytes with pyknotic nucleoli.

# Treatment of ADA Deficiency by BMT

Curative treatment of SCID is possible by BMT patients, and this approach is also being used with success in SCID patients with ADA deficiency (Parkman et al. 1975; O'Reilly et al. 1990). Ideally, such transplants are performed with marrow cells obtained from histocompatible HLA-identical siblings as donors. Under this circumstance, the risk of inducing graft versus host disease is low. It is obvious that the risk of graft rejection, i.e., a host mediated immune response against the graft, is significantly lower in patients with SCID than in other transplant patients with normal immune functions. While in other transplant patients immunosuppressive conditioning by chemotherapy is usually required in order to prevent this complication, BMT in SCID has repeatedly been successful without such conditioning.

## HLA-Identical BMT

An example of the course of an HLA-identical BMT, which we observed in one of our patients, is shown in Fig. 1. As a serious complication, this patient had developed wide spread mycobacterial infection secondary to BCG vaccination as a neonate. Following intravenous infusion of the marrow graft, immune functions developed within a few weeks. The patient mounted a vigorous immune

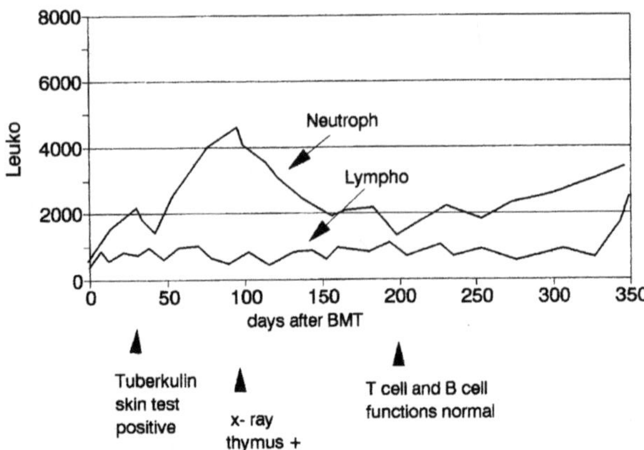



**Fig. 1.** The course of treatment by bone marrow transplantation in an infant with ADA deficiency transplanted from an HLA-identical sister. The patient has now survived for more than 2 years·with normal immune functions and is free of infectious complications

response against her mycobacterial infection, which then caused significantly more complications then prior to BMT with formation of multiple inflammatory abscesses. This complication resolved only gradually over the course of several months. Significant lymphocytopenia persisted for almost 2 years after the BMT and only recently normalized, while both T and B cell functions have remained stable since shortly after BMT.

## HLA-Nonidentical BMT

In the other four patients, no HLA-identical siblings were available. Here, HLA-haploidentical parents were used as marrow donors. In order to prevent the significant risk of graft versus host disease in these transplant patients, additional measures were taken. Mature T cells contained in the marrow grafts were depleted using in vitro purging, an approach successfully applied in other patients with SCID and normal ADA (Friedrich et at. 1984). The course of BMT in our initial ADA deficient patient, transplanted from his HLA-haploidentical father, is shown in Fig. 2. Unexpectedly and in contrast to other SCID patients, this transplant did not lead to immune reconstitution; there was no evidence of engraftment by donor cells even as late as 6 months after BMT and following infusion of a second marrow graft from the father. Although immune functions had never been demonstrable in the patient, we could not completely rule out the possibility of graft rejection. Based on such a presumption, the patient received immunosuppressive conditioning by chemotherapy prior to repeated BMT. Following this procedure, prompt engraftment and full immunological reconstitution developed. The patient could be discharged several months later. He has remained immunologically normal ever since and is now 9 years of age. For unknown reasons, his psychomotor development has remained mildly delayed. Similar experiences of graft failures have been reported by other centers

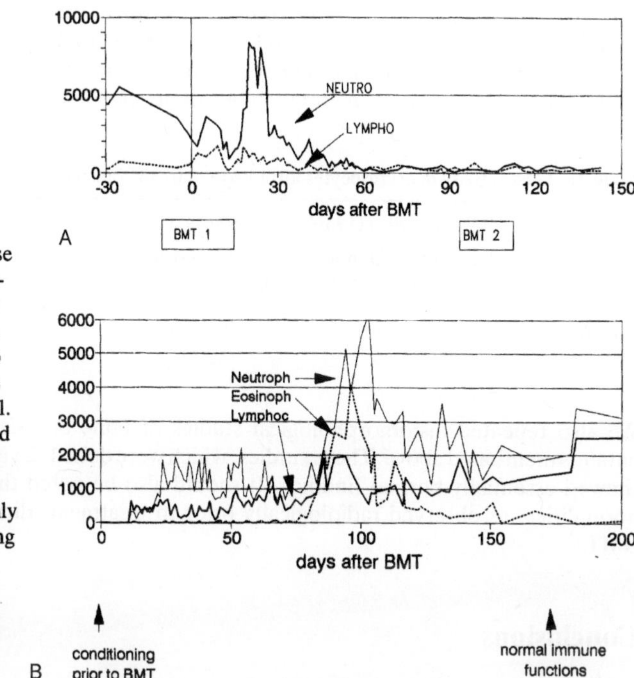



**Fig. 2.** The course of HLA-nonidentical BMT in an infant with ADA deficiency. **A** Two initial transplants were unsuccessful. **B** Engaftment and reconstitution of normal immune functions were only observed following cytoreductive immunosuppressive conditioning prior to the third transplant

and we, like most other transplant centers treating ADA deficient patients, are using protocols now which have incorporated the use of immunosuppressive conditioning prior to HLA-nonidentical T cell depleted BMT. In two of the other three patients, this approach was also successful and now, several years after BMT, they have normal immunity and are free of infectious complications. Both these children show normal mental development. In one patient, the transplant course unfortunately was complicated by severe and lethal graft versus host disease related to transfusion of insufficiently irradiated blood products prior to BMT.

## Enzymatic Findings After BMT

It is important to note that effective treatment of SCID by BMT is not necessarily associated with complete replacement of the lymphohematopoietic system by cells of the donor. More commonly, in particular when no conditioning treatment is used prior to BMT, mixed chimerism will result, with only T cells exclusively of donor origin but with most other blood cells remaining of host type. Similar findings in regard to chimerism have been observed in patients with ADA deficiency. It does not come as a surprise that determinations of enzyme activities after successful BMT vary according to the cell type analyzed and the degree of chimerism obtained. In table 2, findings of repeat enzyme studies after BMT are demonstrated in two of our patients.

**Table 2.** ADA activity in erythrocytes after bone marrow transplantation

|           | ADA activity (U/I EC) | | Red cell | Type of |
|           | Patient 1 | Donor | chimerism | BMT |
|-----------|-----------|-------|-----------|-----|
| Patient 1 | 201       | 192   | Donor 95% | HLA-haploidentical |
| Patient 5 | 3.0       | 201   | Host 100% | HLA-identical |

We also repeated histomorphological studies of bone biopsies in the three patients mentioned above. These studies, which were done 1–2 years after BMT, showed essentially normal findings. It should also be noted that skeletal abnormalities, as observed radiologically prior to treatment, disappeared after BMT.


## Conclusions

Observations in the patients presented here demonstrate well the most prominent clinical, immunological, and pathological abnormalities associated with SCID and profound ADA deficiency. Surprisingly, within the patient population referred to in our center over a time period of 10 years, less severe forms of the disease or variants characterized by delayed onset were not observed.

Treatment of ADA deficient SCID patients by BMT is effective in completely and permanently reversing the life threatening immunological deficiencies related to the underlying metabolic abnormalities. In addition, other clinical findings such as skeletal abnormalities are reversible. This outcome is observed even when reconstitution by donor derived cells remains confined to T cells, as commonly observed after HLA-identical BMT. Obviously, this treatment not only leads to the establishment of an operationally functional T cell system. In addition, engrafted cells also serve as a site to take up and degrade accumulating metabolites toxic to other cell systems, which leads to a reversal of the abnormal functions of these cell systems. This "filter" function of donor derived, enzymatically normal cells is also born out by observations made by several investigators, who showed a marked decrease of measurable plasma metabolite levels after BMT. The successful application of BMT in ADA deficient patients thus serves to demonstrate an important potential of this treatment, which is currently being further explored in the treatment of other metabolic disorders by BMT.

*Acknowledgements.* The dedicated patient care of the nursing staff of the transplant unit is gratefully acknowledged. Histomorphological studies were performed by Prof. H. Stoess, University of Erlangen, Erlangen, FRG.

# References


1. Cederbaum SD, Kaitila I, Rimoin DL, Stiehm ER (1976) The chondro-osseous dysplasia of adenosine deaminase deficiency with severe combined immunodeficiency. J Pediatr 89: 737–742
2. Cohen A, Hirschhorn R, Horowitz SD, Rubinstein A, Polmar SH, Hong R, Martin DW (1978) Deoxyadenosine triphosphate as a potentially toxic metabolite in adenosine deaminase deficiency. Proc Natl Acad Sci USA 75: 472–476
3. Friedrich W, Goldman SF, Vetter U, Fliedner TM, Heymer B, Peter HH, Reisner Y, Kleinhauer E (1984) Immunological reconstitution in severe combined immunodeficiency after transplantation of HLA haploidentical, T cell depleted bone-marrow. Lancet i: 761–764
4. Giblett ER, Anderson JE, Cohen F, Pollara B, Meuwissen HJ (1972) Adenosine-deaminase deficiency in two patients with severely impaired cellular immunity. Lancet 2: 1067–1069
5. Giblett ER, Ammann AJ, Wara DW, Sandman R, Diamond LK (1975) Nucleoside phosphorylase deficiency in a child with severely defictive T-cell immunity and normal B-cell immunity. Lancet 1: 1010–1013
6. Hirschhorn R (1979) Clinical delineation of adenosine d deaminase deficiency. In: Enzyme defects and immune dysfunction, Ciba Foundation symposium 68. Excerpta Medica, Amsterdam, pp 35–49
7. Hirschhorn R, Vawter GE, Kirkpatrick JA Jr, Rosen FS (1979) Adenosine deaminase deficiency: frequency and comparative pathology in autosomally recessive severe combined immunodeficiency. Clin Immunol Immunopathol 14: 107–120
8. O'Reilly RJ, Keever CA, Small TN, Brochstein J (1990) The use of HLA-nonidentical T-cell depleted marrow transplants for correction of severe combined immunodeficiency disease. Immunodeficiency Rev 1: 273–309
9. Parkmann R, Gelfand EW, Rosen FS, Sanderson A, Hirschhorn R (1975) Severe combined immunodeficiency and adenosine deaminase deficiency. N Engl J Med 292: 714–719
10. Rosen FS, Cooper MD, Wegewood RJP (1984) The primary immunodeficiencies. N Eng J Med 311: 235–242
11. Wolfson JJ, Cross VF (1975) The radiographic findings in 49 patients with combined immunodeficiency. In: Meuwissen JH, Pickering RJ, Pollara B, Porter IH (eds) Combined immunodeficiency disease and adenosine deaminase deficiency: a molecular defect. Academic New York; p 255

# 3 The Biochemical Basis and Pathophysiology of ADA and PNP Deficiencies

W. Gutensohn

## Introduction

In the majority of inherited disease we are faced with the following situation: we usually know the clinical phenotype – including even its variations – and the mode of inheritance. In many cases, we also have identified the relevant genes and maybe even the mutations in these genes. With the advent of molecular methods in human genetics, knowledge about specific genes and their mutations is rapidly growing. Nonetheless, the relationship between the genotype and the phenotype more often than not is shrouded in complete ignorance. For example in many diseases we do not understand the mechanism leading from the geno-type to the phenotype nor do we know anything about the causes of variation within the clinical picture.

Strangely enough, in the case of adenosine deaminase (ADA) and purine-nucleoside phosphorylase (PNP) deficiencies, we find ourselves in quite the opposite situation. Within the last 15 years so many different mechanisms explaining the pathophysiology leading to the resulting immunodeficiencies have been proposed and experimentally verified that we now have to sort out those mechanisms which are relevant in the in vivo situation (Kredich and Hersfield 1989).

## Properties of Normal Human ADA and PNP

Both ADA and PNP are intracellular, soluble, cytosolic enzymes. As can be seen from the scheme in Fig.1 the reactions catalyzed by ADA and PNP are embedded in the intermediate metabolism of purines. The steps are truly "intermediate" in the sense that the products of the consecutive reactions (hypoxanthine or guanine) can either be channeled back into the purine nucleotide pool via salvage synthesis or – depending on the particular tissue – go on to final catabolism to uric acid. Not included in the scheme is the fact that both enzymes

---

*Abbreviations:* ADA adenosine deaminase, PNP purine nucleoside phosphorylase, Ado adenosine, dAdo deoxyadenosine, Guo guanosine, dGuo deoxyguanosine, Ino inosine, dIno deoxyinosine, G guanine, H hypoxanthine, SAM *S*-adenosyl-methionine, SAH *S*-adenosyl-homocysteine

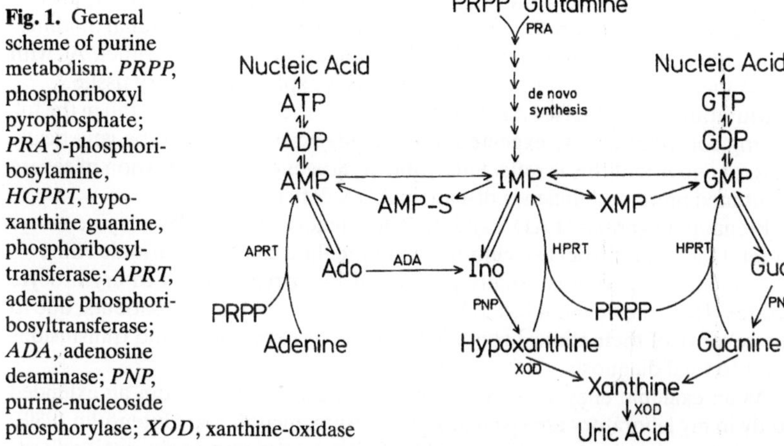



**Fig. 1.** General scheme of purine metabolism. *PRPP*, phosphoriboxyl pyrophosphate; *PRA* 5-phosphoribosylamine, *HGPRT*, hypoxanthine guanine, phosphoribosyltransferase; *APRT*, adenine phosphoribosyltransferase; *ADA*, adenosine deaminase; *PNP*, purine-nucleoside phosphorylase; *XOD*, xanthine-oxidase

also accept the respective deoxynucleosides as substrates and this is important for the pathophysiology (see below). PNP does not act on adenosine. The equilibrium of the PNP reaction is in the direction of nucleoside synthesis. Nevertheless in vivo phosphorolysis is predominant due to high intracellular concentrations of inorganic phosphate.

Normal ADA, as derived from the cDNA and genomic sequences (Wiginton et al. 1986), is a single polypeptide composed of 363 amino acid residues (362 in the mature form) with a calculated molecular weight of approximately 41K. Physically it appears in many different forms. This variation is caused by: (1) genetically determined polymorphism. There are two frequent alleles, designated *ADA¹* and *ADA²*, codominantly expressed and leading to three phenotypes, the homozygous ADA1 and ADA2 and the heterozygous *ADA21*. The frequency of the rare allele, *ADA²*, ranges from 0.03–0.11 in different human populations. Other rare variants have been described. The occasional occurrence of a null allele points to a heterozygous carrier state for the deficiency. (2) The catalytic unit of the enzyme binds to a "complexing protein". This is a glycoprotein, a dimer in the native state, with a molecular weight of 190 K. One such dimer can bind two catalytic units of ADA. Variation in the carbohydrate moiety of the complexing protein produces many different electrophoretic forms of the total complex. The physiological function of the complexing protein is largely unknown. The catalytic unit of ADA is expressed from the same gene in every tissue, thus a defect of ADA is seen in all tissues.

Normal PNP in the native state is a homotrimer with a molecular weight of 87 K–94 K. The subunit – again derived from the cDNA and genomic (Williams et al. 1984) structure – is composed of 289 amino acid residues with a calculated molecular weight of approximately 32 K. PNP also appears in many different physical forms. Here, however, no genetically determined polymorphism is known. All electrophoretic variation is produced by posttranslational modifications, multiplied by the possibilities of combination within the trimer. As with ADA there is just one gene for PNP and this is expressed in every tissue.

Mutations at the loci for ADA and PNP are described in more detail in the following contribution. As expected, both deficiencies turn out to be genetically heterogenous and this is reflected by the heterogeneity of expression of severe combined immunodeficiency disease in ADA deficiency.

Prenatal diagnoses of ADA and PNP deficiency is possible by enzyme assays of fetal blood erythrocytes, cultured amniotic fluid cells, or chorionic villi.

Postnatal diagnosis is usually performed by enzyme assays in erythrocyte lysates. Here, however, it has to be taken into account that the patients, due to the severity of their clinical condition, might have received blood transfusions prior to final diagnosis.

As an example enzyme activities in normal chorionic villi from an extended study in my laboratory are given in Table 1 (Burgemeister and Gutensohn 1989). It is obvious that the normal ranges of both enzyme activities in chorionic villi are rather wide so that on this basis, because of wide overlap, diagnosis of a heterozygous carrier state is not feasible. This has been stated by a number of laboratories.

**Table 1.** Enzyme activities in chorionic villi derived from the 7th–12th week of gestation of normal pregnancies

| Enzyme | $n$ | Mean +/− S.D. (nmol/h · mg protein) | Range (nmol/h · mg protein) |
|---|---|---|---|
| ADA | 111 | 1292.6 +/− 347.8 | 726.0– 2202.0 |
| PNP | 110 | 7377.9 +/− 3186.4 | 1335.0–14047.0 |

ADA, adenosine deaminase; PNP, purine-nucleoside phosphorylase

## Role of ADA and PNP in Normal Purine Metabolism

Both ADA and PNP have purine nucleosides as their substrates; thus in the following we will concentrate on nucleosides. First some general facts can be stated:

– Concentrations of adenosine in plasma are low, the concentrations of all the other purine nucleosides, including the deoxynucleosides, are usually reported as "undetectable".
– Although concentrations of nucleosides in plasma are rather low, nucleoside fluxes through plasma are estimated as being high.
– There is a rapid equlibration of nucleosides between the extra- and the intracellular space, brought about by carrier-mediated facilitated diffusion.

Sources of purine ribonucleosides, especially adenosine include transient intermediates of normal nucleic acid turnover and byproducts of *S*-adenosyl-methionine (SAM) dependent transmethylation reactions. In this way 14–23 mmol of adenosine are formed per day from $S$-adenosyl-homocystein (SAH) (Fig. 4) Hepatic synthesis of creatine alone amounts to about 85% of this. Adenosine may also be released from certain tissues. Finally, a major extracellular source of adenosine may be the breakdown of nucleotides from senescent and dying hematopoietic cells, catalyzed by ectonucleotidase cascades on the surface of cells, e.g., vascular endothelium.

However, it should be stressed that the main route of intracellular adenosine monophosphate (AMP) catabolism is via its deamination to inosine monophosphate (IMP) and this is catalyzed by adenylate deaminase and not ADA. The purine nucleotide cycle, which is important for muscle tissue and which will be dealt with in detail in other chapters of this book, does not include either ADA or PNP catalyzed steps.

In contrast to the ribonucleosides, which may also appear in pathways not related to RNA synthesis or breakdown, deoxynucleosides only play a role in the context of DNA metabolism: (1) as constituents of substrate cycles which serve the regulation of deoxynucleoside triphosphate pools. (2) as intermediates in DNA degradation. In contrast to the ribonucleotides, e.g., AMP, catabolism of deoxyAMP (dAMP) and deoxyguanosine monophosphate (dGMP) proceeds almost exclusively via dephosphorylation to deoxyadenosine (dAdo) and deoxyguanosine (dGuo), respectively.

Many of the enzymes involved in these intermediate or catabolic routes have been studied extensively with respect to their kinetic properties and substrate specificities. On the basis of these data the predominant metabolic fates of the purine intermediates, given their physiological intracellular concentrations, have been estimated. A crude and maybe simplified summary of these results is given in Table 2, which contrasts the normal situation with what happens in the deficiencies; however it is still rather general. The role of immune competent cells, i.e., the lymphocytes, is discussed below.

**Table 2.** Fates of ribonucleosides and deoxyribonucleosides in normal metabolism: predominant routes

| Reaction | Fate |
| --- | --- |
| Ado → AMP (adenosine kinase) | |
| | Deamination to Ino only at increased Ado concentrations |
| dAdo → dIno (ADA) | |
| | Phosphorylation of dAdo to dAMP only in ADA deficiency |
| Guo, Ino, dGuo, dIno → G or H + Rib-1-P or dRib-1-P (PNP) | |
| | Phosphorylation of dGuo to dGMP only in PNP deficiency |

Ado, adenosine; dAdo, deoxyadenosine; Ino, inosine; dINO, deoxyinosine; ADA, adenosine deaminase; AMP, adenosine monophosphate; Guo, guanosine; dGuo, deoxyguanosine; G, guanine; H, hypoxanthine; Rib-1-P, Ribose-1-phosphate; dRib-1-P, deoxyribose-1-phosphate; PNP, purine-nucleoside phosphorylase

**Table 3.** Plasma levels of nucleosides

|  | Control ($\mu M$) | ADA deficiency ($\mu M$) | PNP deficiency ($\mu M$) |
|---|---|---|---|
| Ado | 0.05 – 0.4 | 0.1 – 10 |  |
| Ino | Undetectable |  | 14 – 115 |
| Guo | Undetectable |  | 6 – 29 |
| dIno | Undetectable |  | 2 – 19 |
| dGuo | Undetectable |  | 2 – 14 |
| Urate | 220 +/− 60 | 80 – 260 | Trace – 150 |

Adapted from Kredich and Hershfield (1989).
ADA, adenosine deaminase; PNP, purine-nucleoside phosphorylase; Ado, adenosine; Ino, inosine; Guo, guanosine; dIno, deoxyinosine; dGuo, deoxyguanosine.

**Table 4.** Urine levels of nucleosides

|  | Control (mmol/mol creatinine) | ADA deficiency (mmol/mol creatinine) |
|---|---|---|
| Ado | <1 | 10.3 +/− 20.4 |
| dAdo | <1 | 140 +/− 74 |
| Uric acid | 400 – 1000 | 460 – 1600 |

Adated from Kredich and Hershfield (1989).
ADA, adenosine deaminase; Ado, adenosine; dAdo, deoxyadenosine.

**Table 5.** Urine levels of nucleosides

|  | Control (mmol/g creatinine) | PNP deficiency (mmol/g creatinine) |
|---|---|---|
| Ino | Undetectable | 4.5 – 17 |
| dIno | Undetectable | 2.8 – 4.7 |
| Guo | Undetectable | 1.4 – 7.7 |
| dGuo | Undetectable | 2.3 – 3.6 |
| Uric acid | 2.8 – 5.2 | 0.16 – 3.13 |

Adapted from Kredich and Hershield (1989).
PNP, purine-nucleoside phosphorylase; Ino, inosine; dIno, deoxyinosine; Guo, guanosine; dGuo, deoxyguanosine.

## This leads us to the question: What Goes Wrong in Metabolism in Either ADA or PNP Deficiency?

The obvious approach to answering this question would be to study the patients themselves, but due to the extreme lymphopenia (especially in ADA deficiency) the cells probably most severely affected cannot be investigated.

What can be and has been done in these patients is an analysis of purine metabolite levels in body fluids, excretions, and erythrocytes. The data are summarized in Tables 3–5. The general conclusions are: (a) Ribo- and deoxyribonucleosides, normally at low to undetectable concentrations, appear in significant amounts in body fluids and urine. (b) Purine production in ADA deficiency is normal. (c) Although PNP deficiency is characterized by hypouricemia and hypouricosuria, overall, purine overproduction is observed, with inosine as the main excretion product.

## Metabolite Levels in Erythrocytes in ADA and PNP Deficiency

In ADA deficiency, deoxyadenosine triphosphate (dATP) concentrations are increase dramatically, from almost undetectable in controls to levels comparable to or higher than those of ATP. These high dATP concentrations are already detectable in fetal blood or cord blood and increase further after birth. In extreme cases this leads to a depletion of erythrocyte ATP pools (for explanation of a mechanism see below).

In PNP deficiency, deoxyguanosine triphosphate (dGTP) levels are increased from undetectable in controls to 2–11 nmol/ml packed cells. At the same time a depletion of GTP pools is observed.

For the reasons mentioned above information about deoxynucleoside triphosphate pool expansions in the lymphoid cells of ADA and PNP deficiency patients is more difficult to get, but some observations have been reported. In PNP deficiency dGTP pool expansion has never been seen in nonerythroid cells. An interesting case is one ADA deficient patient who had undergone a successful transplantation of peripheral blood leukocytes. Here an increase of dATP ($40$–$80$ pmol/$10^6$ cells) was observed in functioning lymphocytes of donor origin (Rich et al. 1980).

The difficulty in studying lymphoid cells of ADA and PNP deficient patients was largely circumvented by the creation of powerful pharmacological models. In tissue culture, ADA in normal cells is inhibited by erythro-9-(2-hydroxy-3-nonyl) adenine (EHNA) with a $K^i$ of $1.6 \cdot 10^{-9}$ $M$, by coformycin with a $K^i$ of $1.0 \times 10^{-11}$ $M$, and by 2-deoxycoformycin with a $K^i$ of $2.5 \cdot 10^{-12}$ $M$. PNP is inhibited by 8-aminoguanosine with a $K^i$ of $1.7 \cdot 10^{-5}$ $M$. In an in vivo model, leukemic patients, e.g. those with T cell leukemia, were treated with 2-deoxycoformycin.

With these model systems one could test the behavior of lymphoid cells when the functions of either ADA or PNP are blocked. The following general observations were made: (1) Inhibition of either ADA or PNP per se is not toxic to cells. (2) Exogenous nucleosides are, however, toxic for ADA-inhibited cells (Ado toxic at $5$–$50\,\mu M$, dAdo toxic at $0.2$–$50\,\mu M$) and PNP-inhibited cells (dGuo toxic at $1$–$50\,\mu M$).

As shown in Table 3 there is an exogenous source of nucleosides in the patients, but the question is why the immune system, i.e., lymphoid cells, should be particularly sensitive to this type of toxicity. Early investigations of this problem seemed to yield a rather clear-cut picture. The hypothesis was that lymphoid cells, due to their specific enzyme makeup, were doomed to take up and accumulate all the excess ribo- and deoxyribonucleosides produced by all tissues of the enzyme deficient patients and distributed via the blood stream. However, by extended studies it turned out that the situation is not quite as simple. A rather crude overview of the behavior of different human lymphoid cell types is given in Table 6.

**Table 6.** Capacity to accumulate deoxynucleoside triphosphates in different lymphoid cell types

| Lymphoid cell type | dATP accumulation (from dAdo) | dGTP accumulation (from dGuo) | Sensitivity to Ado and dAdo in colony assay when ADA inhibited |
|---|---|---|---|
| Lymphoblastoid cell lines | | | |
|   T cell lines | + | + | |
|   B cell lines | +/− | +/− | |
| Malignant lymphoblasts | | | |
|   T ALL | +/− Dependent on maturity | | |
|   Non T ALL | − | | |
| Thymocytes (dividing and nondividing) | + | + | |
| Mature circulating T cells | − | − | High |
| Peripheral blood or tonsil B cells | + | − | High |

dATP, deoxyadenosine triphosphate; dAdo, deoxyadenosine; dGTP, deoxyguanosine triphosphate; dGuo, deoxyguanosine; Ado, adenosine; ADA, adenosine deaminase.

One certainly cannot say that lymphoid cells in general are particularly sensitive to deoxynucleoside toxicity. Nonetheless, there may be certain subpopulations or maturation stages which are extremely sensitive. Furthermore, dATP accumulation is not necessarily restricted to lymphoid tissues. Rather, a somewhat lower susceptibility of some B cell populations might be inferred from these data. This would explain the differences in the clinical phenotype which are evident in PNP deficiency but can also be seen in some ADA deficient patients.

Now, with all that the stage is set for the final question: What are the pathobiochemical mechanisms leading to immune dysfunction in ADA and PNP deficiency? A summary of the major mechanisms that have been proposed is given

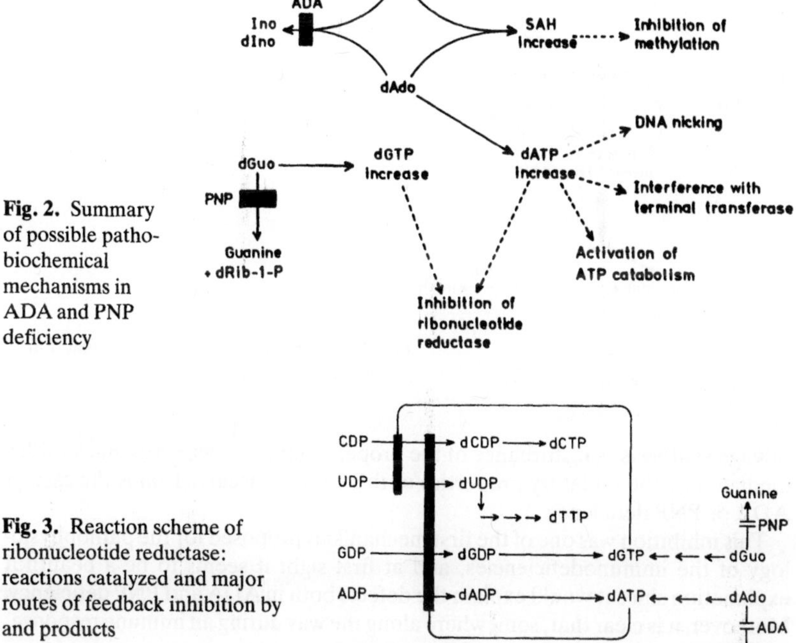


**Fig. 2.** Summary of possible pathobiochemical mechanisms in ADA and PNP deficiency

**Fig. 3.** Reaction scheme of ribonucleotide reductase: reactions catalyzed and major routes of feedback inhibition by and products

in Fig. 2. From this scheme a number of possibilities have been deliberately left out; these were discussed earlier but are now regarded as less likely causes of the disease.

The first relevant mechanism is inhibition of ribonucleoteide reductase. As mentioned above, in contrast to the ribonucleotides, which serve many different functions in the cell, deoxyribonucleotides are exclusively precursors of DNA synthesis. There are no major pools from which synthesis could draw, so the deoxynucleotides must be provided the moment they are needed, i.e., during DNA replication. They must not only be provided in adequate amounts but also in the proper concentration ratios, because otherwise the replication machinery will make mistakes. To guarantee a sufficient and balanced supply of deoxyribonucleotides is the job of the complex enzyme ribonucleotide reductase (Fig. 3).

The following features of this enzyme are important: (1) Reduction occurs at the level of the nucleoside diphosphates. (2) There is just one enzyme for the four different nucleotides. (3) The enzyme is therefore regulated by a complex system of feedback inhibition and activation. A simplified version of the feedback inhibition by the end products dATP and dGTP is given in Fig. 3. (4) Whenever one of the deoxyribonucleoside triphosphates is fed into the system via

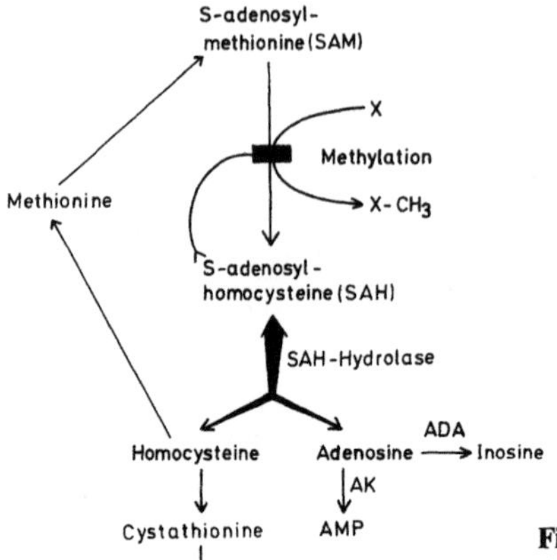


**Fig. 4.** Methylation and sequential reactions; *AK*, adenosine-kinase

salvage synthesis, a disturbance of the proper balance of deoxyribonucleotides mediated by the reulatory properties of this enzyme occurs. That is the case in ADA or PNP deficiency.

This inhibition was one of the first mechanisms proposed for the pathophysiology of the immunodeficiencies, and at first sight it seems to be a beautiful explanation since it would explain the defects both in ADA and PNP deficiency. Moreover, it is clear that, somewhere along the way during an immune response, proliferation of cells is an essential event.

However, experimental evidence was provided for deoxyadenosine toxicity in ADA-inhibited lymphoid cells in situations in which blockade of ribonucleotide reductase cannot be the cause. These include normal circulating peripheral lymphocytes, which are nondividing; early during lymphocyte stimulation, when DNA replication is not yet initiated and ribonucleotide reductase activity is very low; and in situations in which cells do not (see Table 6) or cannot accumulate dATP or dGTP because of mutations. Clearly there must be other mechanisms, as indicated in Fig. 2. One of these is inhibition of methylation. A general scheme of the methylation reactions and the feedback regulation of the system is shown in Fig. 4.

The equilibrium of the reaction catalyzed by (SAH) hydrolase is in the direction of SAH synthesis. To avoid the accumulation of SAH and feedback inhibition of methylation, SAH hydrolysis has to be driven by removal of Ado and homocysteine from the equilibrium by coupled reactions. The ratio of SAM to SAH in cells is called the "methylation index" and is an indication of the cell's capacity to methylate.

Two mechanisms were shown to be relevant for this system in ADA deficiency: (1) An influence of increased Ado concentrations (see Table 3) on the equilibrium of the SAH hydrolase reaction. (2) A "suicide-like" inactivation of SAH hydrolase by dAdo (Hershfield 1979a). Both mechanisms operate in vivo. This has been described in patients with different forms of leukemias treated with deoxycoformycin, including cases in which the malignant T cell accumulate dATP, cases in which malignant T or B cells do not accumulate dATP, and in ADA deficient children (Hershfield 1979b).

The in vivo effects are dramatic decreases of both the methylation index (from 40:1 to 3:1) as observed in circulating lymphoblasts of a leukemic patient treated with deoxycoformycin, and SAH hydrolase activity in erythrocytes of ADA deficient children (to less than 2% of normal).

Given the general importance of methylation, inhibition of this reaction will interfere with many different functions in both resting and dividing cells.

ATP depletion is another mechanism which has been considered and is shown in Fig. 2. The underlying observation was ATP depletion in erythrocytes leading, in extreme cases, to half normal concentrations and hemolytic anemia in both ADA deficient children (Simmonds et al. 1982) and in patients treated with deoxycoformycin (Siaw et al. 1980). Similar effects were seen in circulating T lymphoblasts.

The drop in ATP levels is dependent on dATP accumulation. On a metabolic basis this suggests that dAMP in contrast to AMP, is not a good substrate for AMP deaminase and that dATP, like ATP, is a potent activator of AMP deaminase and cytoplasmic 5'-IMP nucleotidase. Together this results in stimulation of intracellular ATP breakdown.

An alternative explanation for this ATP depletion was given in the context of DNA nicking (see Fig. 2). This was based on the following observations: (1) Resting lymphocytes normally contain strand breaks which are rejoined upon mitogenic stimulation. (2) Treatment of resting lymphocytes with dAdo induces strand breaks and, as a consequence, depletion of nicotinamide-adenine dinucleotide (NAD) and ATP pools (Seto et al. 1985).

From this a hypothetical sequence of events was postulated: dATP interferes in some way with DNA repair → accumulation of strand breaks → induction of polyADP-ribose synthetase (uses NAD as substrate) → depletion of NAD pools → depletion of ATP pools → cell death. This hypothesis has not yet been verified by observations in either patients treated with deoxycoformycin or ADA deficient children.

Further mechanisms, as listed below, have only been shown in in vitro systems so that we can only speculate about their possible relevance for the pathophysiology in ADA and PNP deficiency. Such mechanisms include:

- Increases in dATP concentrations could change the product of the reaction of terminal deoxynucleotidyl transferase (TdT), a template independent DNA polymerase. TdT has a possible role in the generation of immunological diversity.
- dATP in the presence of $Mn^{2+}$ inhibits TdT.
- dAdo inhibits RNA synthesis in ADA-inhibited blood lymphocytes.
- SAH inhibits initiation of transcription by RNA polymerase II.
- SAH inhibits mRNA cap methylation.

- Ado induces expression of CD8 antigen and suppressor activity in previously CD4$^+$ T helper cells (via adenosine receptor?).
- dAdo inhibits the function of T cells in mixed lymphocyte culture (MLC) and expression of interleukin-2 (IL-2) and/or the IL-2 receptor.
- dGuo blocks antigen-induced T suppressor cell activity in vitro and development of suppressor T cells in vivo (in mice).

A number of observations point to a specific interference with normal lymphoid differentiation by dAdo, when ADA is inhibited (by deoxycoformycin): First, committed progenitor cells, derived from either bone marrow (Aye et al. 1982) or peripheral blood (Brox et al. 1982), cannot develop into erythroid, myeloid, or T lymphoid colonies. This holds even for conditions in which survival of the marrow cells is enhanced by spleen cell conditioned medium. Second, there are indications that a special requirement for ADA activity exists at the earliest stage of T cell differentiation (Shore et al. 1981). Third, abrupt conversion from a T lymphoblastoid to a myeloid phenotype of the malignant cells was reported in two patients with a diagnosis of stem cell leukemia who were under therapy with deoxycoformycin. These results were corroborated by studies in vitro with leukemic cells of one of the patients and with a cell line derived from them (Hershfield et al. 1984). Such an interference with lymphoid differentiation would correspond to the absence of lymphoid lineages in ADA deficient children.

## Concluding Remarks

This short review has considered a confusing multitude of possible explanations for the pathophysiology of ADA and PNP deficiency. Can we finally decide which of these mechanisms is relevant in vivo? At the present state of our knowledge the answer to this question must be "no". However, there are a number of facts which can be stated with reasonable confidence: For example, it is certainly not justified to reduce the explanation of the pathophysiology to one single cause, e.g., the inhibition of ribonucleotide reductase. Several of the other mechanisms mentioned, most probably acting in parallel, must be considered. In addition, at certain, probably early, stages of differentiation lymphoid cells may be particularly dependent on ADA and PNP for protection against the accumulation of toxic metabolites, since in their environment, e.g., the thymus, programmed cell death on a massive scale is a physiological event. Finally, a specialty of lymphoid cells is gene rearrangement and somatic mutation, creating the diversity of the immunological repertoire. Interference at any one of these steps by unphysiological concentrations of deoxyribosyl metabolites would represent a very plausible reason for the disturbance of a function unique to the immune system. However, the details remain to be elucidated by further experimentation.

# References

Aye MT, Dunn JV, Yang WC (1982) Studies on the effect of deoxyadenosine on deoxycoformycin-treated myeloid and lymphoid stem cells. Blood 60: 872

Brox LW, Pollock E, Belch A (1982) Adenosine and deoxyadenosine toxicity in colony assay systems for human T-lymphocytes, B-lymphocytes, and granulocytes. Cancer Chemother Pharmacol 9: 49

Burgemeister R, Gutensohn W (1989) Normal ranges of the activities of the different enzymes in 100 independent preparations of chorionic villi. Comparison of specimens from induced abortions, biopsies, and cultured cells. Prenatal Diagnosis 9: 195–204

Hershfield MS (1979a) Apparent suicide inactivation of human lymphoblast S-adenosylhomocysteine hydrolase by 2'-deoxyadenosine and adenosine. J Biol Chem 254: 22–25

Hereshfield MS, Kredich NM, Ownby H, Buckley R (1979b) In vivo inactivation of erythrocyte S-adenosylhomocysteine hydrolase by 2'-deoxyadenosine in adenosine deaminase-deficient patients. J Clin Invest 63: 807–811

Hershfield MS, Kurtzberg J, Moore JO, Whang-Peng J, Haynes BF (1984) Conversion of a stem cell leukemia from T-lymphoid to a myeloid phenotype by the adenosine deaminase inhibitor 2'-deoxycoformycin. Proc Natl Acad Sci USA 81: 253–257

Kredich NM and Hershfield MS (1989) Immunodeficiency diseases caused by adenosine deaminase deficiency and purine nucleoside phosphorylase deficiency. In: Scriver CM, Beaudet AL, Sly W, Valle D (eds) The metabolic basis of inherited disease, 5th edn. McGraw-Hill, New York, p 1045–1075

Rich KC, Richman CM, Meijas E, Daddona PA (1980) Immunoreconstitution by peripheral blood leukocytes in adenosine deaminase-deficient severe combined immunodeficiency. J Clin Invest 66: 389–395

Seto S, Carrera CJ, Kubota M, Wasson DB, Carson DA (1985) Mechanism of deoxyadenosine and 2-chlorodeoxyadenosine toxicity to nondividing human lymphocytes. J Clin Invest 75: 377–383

Shore A, Dosch HM, Gelfand EW (1981) Role of adenosine deaminase in the early stages of precursor T cell maturation. Clin exp Immunol 444: 152–155

Siaw MFE, Mitchell BS, Koller CA, Coleman MS, Hutton JJ (1980) ATP depletion as a consequence of adenosine deaminase inhibition in man. Proc Natl Acad Sci USA 77: 6157–6161

Simmonds HA, Levinsky RJ, Perret D, Webster DR (1982) Reciprocal relationship between erythrocyte ATP and deoxy-ATP levels in inherited ADA deficiency. Biochem Pharmacol 31: 947–951

Wiginton A, Kaplan DJ, States JC, Akeson AL, Perme CM, Bilyk IJ, Vaughn AJ, Lattier DL, Hutton JJ (1986) Complete sequence and structure of the gene for human adenosine deaminase. Biochemistry 25: 8234–8244

Williams SR, Goddard JM, Martin DW Jr (1984) Human purine nucleoside phosphorylase cDNA sequence and genomic clone characterization. Nucl Acid Res 12: 5779–5787

# 4 The Genetic and Metabolic Basis of ADA Deficiency

M. S. Hershfield, F. X. Arredondo-Vega, I. Santisteban, and S. Chaffee

## Introduction

It is nearly 20 years since the first report of adenosine deaminase (ADA) deficiency occurring with severe combined immunodeficiency disease (SCID) [1]. At the time, research on purine metabolism in human disease was focused on questions related to gout. Investigation of the Lesch-Nyhan syndrome and phosphoribosylpyrophosphate (PPRP) synthetase overactivity fit this paradigm well, providing insight into mechanisms responsible for causing purine overproduction and hyperuricemia. The discovery of an association between ADA deficiency and SCID was entirely unexpected and identified an apparently important, but enigmatic, role of purine nucleoside metabolism in the development and function of the immune system. It also highlighted a lack of understanding of the biochemical control of the immune system at a time when the existence and functions of distinct lymphocyte subsets were only beginning to be appreciated. Giblett's report [1] and her subsequent discovery of purine-nucleoside phosphorylase (PNP) deficiency with selective T cell dysfunction offered a novel and challenging direction for future research.

The first decade of research on ADA and PNP deficiencies explored the metabolism and cytotoxic effects of purine nucleosides that are responsible for selective immune dysfunction in ADA and PNP deficiencies, an area we have previously reviewed [2], and which Dr. Gutensohn has summarized elsewhere in this volume. We will first try to give a necessarily brief overview of research on ADA deficiency conducted during the past 10 years, which has focused on establishing the cause of the enzyme deficiency itself at the molecular level. We will then discuss recent clinical research aimed at developing a specific therapy for ADA deficiency.

## Genetic and Molecular Factors

### ADA Gene, cDNA, and Crystal Structure

Between 1983 and 1986 several groups of investigators reported the isolation and sequencing of cDNA for human ADA [3–7] and PNP [8, 9]. Wiginton, Hutton, and their collaborators then succeeded in cloning the entire ADA gene and determined its structure and complete nucleotide sequence [10]. This body of research (and related investigation of the murine ADA gene by Kellems' laboratory) have made possible important ongoing studies of ADA gene expression and the elucidation of mutations responsible for causing ADA and PNP deficiencies. Then three-dimensional structures of both enzymes have now been determined, providing the opportunity to understand the direct consequences of these mutations on enzyme stability and function [11, 12].

Human ADA is encoded by approximately 32 000 base pairs (bp) of DNA at a single locus on the long arm of chromosome 20 that has been localized to 20q13.11 [13]. The approximately 1.5 kb ADA mRNA is comprised of 12 exons and contains about 95 bp of 5' and 314 of 3'untranslated sequence flanking a 1089 bp open reading frame that encodes a protein of 363 amino acids of calculated Mr 40 762 (the N-terminal methionine appears to be removed by post-translational modification). The mature ADA protein is active as a 40K monomer in erythrocytes and lymphocytes, but in some other tissues the enzyme is combined with a 220kd "ADA complexing protein", the functional significance of which remains uncertain. Comparison of the amino acid sequences of bacterial and mammalian ADA [14] and determination of the three-dimensional structure of recombinant murine ADA [12] have shed light on the evolution of ADA and its catalytic mechanism. The enzyme was shown to have an eight-stranded parallel $\alpha/\beta$ barrel architecture with a very tightly bound zinc atom at the active site (the cofactor requirement had not previously been suspected since purified ADA is fully active after prolonged dialysis). Adenosine monophosphate (AMP) deaminase is also a zinc enzyme and has a reaction mechanism similar to that of ADA; conserved active site amino acid residues present in bacterial and mammalian ADA are also found in AMP deaminase, confirming a common origin of the genes for these enzymes.

## Mutant ADA Alleles in SCID Patients

Several laboratories have analyzed mutant ADA alleles in SCID patients (summarized in Table 1). Two children, one in the Netherlands and the other in the USA, are homozygous for a 3240 bp deletion encompassing the ADA promoter and exon 1, which appears to have arisen from illegitimate recombination between two *Alu*I repeats [15, 16]. A splicing defect, resulting in skipping of exon 4, has been identified in one patient [17]. The majority of patients, however, are compound heterozygotes for point mutations that cause amino acid substitutions [17–22]. This heterogeneity provides the basis for varying degrees of residual catalytic activity and undoubtedly contributes to clinical heterogeneity. The most apparent examples of this are immunologically normal individuals with so-called "partial" ADA deficiency, whose erythrocytes lack ADA activity but whose nucleated cells express approximately 5%–70% of normal activity (compared with 0%–2% for SCID patients). The first "partials" were discovered among the !Kung tribe in Southern Africa, but the majority of the

**Table 1.** Mutations identified in ADA alleles of SCID patients

| Exon | Base change | Mutation | References |
|---|---|---|---|
| 1 | 3250 bp deletion | | |
| 4 | AG > GG 3'splice junction IVS 3 | Missing exon 4 | [17] |
| 4 | AAA > AGA | $Lys_{80} > Arg$ | [22] |
| 4 | CGG > TGG | $Arg_{101} > Trp$ | [17] |
| 4 | CGG > CAG | $Arg_{101} > Gln$ | [19] |
| 4 | CTG > CCG | $Leu_{107} > Pro$ | [24] |
| 7 | CGT > CAT | $Arg_{211} > His$ | [17, 18] |
| 7 | GGG > AGG | $Gly_{216} > Arg$ | [20; authors, unpublished] |
| 10 | CTG > CGG | $Leu_{304} > Arg$ | [22] |
| 11 | GCG > GTG | $Ala_{329} > Val$ | [17] |

ADA, adenosine deaminase; SCID, severe combined immunodeficiency disease; bp, base pair; IVS, intervening sequence.

dozen or so patients were identified in the course of screening newborns for ADA deficiency in New York state in the last 1970s. The latter have been found to be genetically heterogenous; at least seven different alleles have been identified, which for the most part are distinct form those found in SCID patients [23, 24].

Knowledge of the three-dimensional structure will allow a more precise understanding of how specific mutations produce enzyme deficiency. The amino acid substitutions identified thus far do not involve residues that directly contact the zinc cofactor or bound substrate, but some would probably alter the positioning of active site functional groups to a degree that would interfere with the ADA catalytic mechanism [12]. Substitutions distant from the active site may alter enzyme folding to decrease stability or increase susceptibility to intracellular degradation. For example, the $Pro_{297} > Gln$ substitution in a patient with partial ADA deficiency produces a thermolabile enzyme [23]. We have recently described an unusual case in which a point mutation also appears to affect ADA activity in a conditional manner, but depending on cell type rather than temperature [25]. The patient had typical SCID, with high levels of erythrocyte dATP and <1% of normal ADA activity in her erythrocytes, granulocytes, and cultured B cell lines. Paradoxically, interleukin-2 (IL-2) dependent T cells cultured from the patient expressed variable ADA activity, reaching into the normal range, with normal $K_m$, heat stability and sensitivity to ADA inhibitors. The steady state level of ADA mRNA was not elevated in $ADA^+$ T cells and PCR-amplified cDNA from these cells contained the same point mutations that had previously been identified in her ADA- B cells by Akeson et al. [17, 18]. We have postulated that one of these point mutations ($Arg_{101} > Trp$, $Arg_{211} > His$) may be at a site recognized by a proteolytic process expressed in B cells and other hematopoietic cells but not in activated T cells.

# Polyethylene Glycol (PEG-) ADA Replacement Therapy

Techniques for performing T cell-depleted, haploidentical, marrow transplantation were developed in the early 1980s. Prior to that time, a form of replacement therapy, repeated transfusion of irradiated normal red cells, was employed as therapy for patients who lacked an HLA-identical sibling donor [26, 27]. This was feasible because ADA substrates in plasma could enter red cells via the cell membrane associated nucleoside transporter and because dATP in lymphocytes is convertible by metabolism to deoxyadenosine (dAdo), which is in rapid, transporter mediated equilibrium with plasma dAdo. Thus elimination of circulating dAdo can theoretically lead to lowering the level of dAdo and dATP in ADA deficient cells. Although transfused red cells provide a long-circulation reservoir of ADA activity, transfusion therapy is limited by the low ADA activity of erythrocytes, the volume of red cells that can safely be administered, and, in the long run, by iron overload and the risk of transmitting viral disease. Most patients did not respond; few children survived beyond early childhood on transfusion therapy [28], and they remained severely lymphopenic with minimal in vitro lymphocyte function (unpublished observations).

With the advent of haploidentical marrow transplantation, transfusion therapy was for the most part abandoned. However, the significant failure rate and mortality associated with the procedure in many centers [29, 30] has maintained a need to develop alternative approaches to therapy. Gene replacement therapy for ADA deficiency became feasible in 1984–1985, soon after the cloning of the ADA cDNA and following the development of suitable retroviral vectors. However, since achievement of this goal did not appear to be imminent we chose to investigate a novel approach to enzyme replacement.

During the 1970s many techniques were investigated for stabilizing purified enzymes so that they could be used for therapy [31]. Among these was covalent modification with polyethylene glycol (PEG), which both prolonged their circulating life and reduced their immunogenicity [32, 33]. PEG-modified bovine ADA was prepared and proposed as a treatment for ADA deficiency as early as 1981 [34]. However, clinical trial was not attempted for several reasons, including concerns about potential immunogenicity, the unproved safety of PEG-modified proteins (which had never been used in humans), the limited success with transfusion therapy for ADA deficiency, and because effective chronic enzyme replacement therapy had not yet been developed for any metabolic disease.

In late 1985 we reviewed the preclinical studies of PEG-ADA and felt that it could potentially provide much higher levels of circulating ADA activity than red cell transfusion and it appeared to be safe in animal studies. With our colleagues in the pediatric immunology division at Duke University we decided to investigate the use of PEG-ADA (manufactured by Enzo Inc. South Plainfield, NJ, USA) to treat a child who had failed to engraft after haploidentical marrow transplantation and who had not responded to transfusion therapy. Treatment was begun in April, 1986; a second child who had been treated with red cell transfusion for over 8 years began treatment 5 months later. We established a dose that would maintain plasma ADA levels sufficiently high to correct metabolic abnormalities in red cells. In each patient, this was followed by an increase in lymphocyte counts and in vitro lymphocyte responses to mitogens [35].

Since 1986, our laboratory has collaborated with physicians and immunologists at many institutions in the USA, Canada, and Europe to evaluate PEG-ADA therapy in 24 other patients (to date). We have monitored the metabolic effects of PEG-ADA and the development of antibodies to the modified bovine enzyme, while our collaborators have monitored clinical and immunological response to treatment. Results with several patients have been published and we have summarized the experience through 1990, when PEG-ADA was approved for treatment of ADA deficiency by the Food and Drug Administration [35–38]. A report on the antibody response to PEG-ADA (Chaffee et al., unpublished data) has recently been submitted for publication.

In brief, PEG-ADA administered once or twice weekly by intramuscular injection maintains plasma ADA activity at from approximately 1.5 to five times the normal blood (erythrocyte) ADA activity, depending on dose (15–60 units, measured at 25 °C, per kg per week). This reduces the concentration of total dAdo nucleotides (dAXP) in erythrocytes by roughly 100-fold, to within the range observed in healthy individuals with partial ADA deficiency. Antibody to PEG-ADA becomes detectable within 3–8 months of treatment in most patients, but there have been no allergic reactions to PEG-ADA and, except in two patients, there has been no effect on enzyme activity. In one of the latter, tolerance to PEG-ADA could be induced, and in the other enhanced clearance could be overcome by twice weekly dosing (Chaffee et al., submitted for publication).

In the majority of patients treated for more than 6 months, an increase in T cell and total lymphocyte counts and improvement of in vitro lymphocyte response to mitogens follows correction of metabolic abnormalities. Specific in vitro proliferative responses to antigens and specific antibody responses to immunization have been observed in approximately half, and about half the patients treated for longer than a year have discontinued treatment with intravenous immunoglobulin. Most patients remain lymphopenic and in vitro lymphocyte function fluctuates for reasons that are unclear and unrelated to plasma ADA or red cell dAXP levels. About a fifth of patients show minimal recovery of immune function. However, all patients treated for longer than 6 months have shown marked clinical improvement and have remained free of life threatening infections. There have been two deaths among the 26 patients treated to date: a critically ill infant died within a week of starting therapy, and a second infant, who required aggressive treatment with immunosuppressive drugs for severe autoimmune hemolytic anemia, died at 4 months (unpublished results). A recent review of bone marrow transplantation for SCID in Europe reported that 9 of the 19 ADA deficient patients treated by haploidentical bone marrow transplantation had died by 2 years post transplantation [29].

Prior to the development of PEG-ADA replacement therapy, ADA deficient patients either died or were successfully transplanted. The patients undergoing treatment with PEG-ADA provide a unique opportunity to expand our knowledge of the clinical spectrum of the disease and, by following the reconstitution of autologous immune function, to better define the nature and reversibility of the block in immune development caused by ADA deficiency. The basis for variable recovery of immune function during PEG-ADA therapy is unclear at present. It does not appear related to the effectiveness of treatment in maintaining stable circulating levels of ADA activity or in correcting metabolic abnormalities, at least those that can be measured in erythrocytes. It is possible that the same genetic and molecular factors that in part determine clinical severity also determine the capacity for immune reconstitution once treatment is begun. We are presently analyzing the mutations of patients undergoing treatment with PEG-ADA with the aim of establishing the relationship between genotype and phenotype.

# References


1. Giblett ER, Anderson JE, Cohen F, Pollara B, Meuwissen HJ (1972) Adenosine deaminase deficiency in two patients with severely impaired cellular immunity. Lancet 11: 1067–1069
2. Kredich NM, Hershfield MS (1989) Immunodeficiency disease caused by adenosine deaminase deficieny and purine nucleoside phosphorylase deficieny. In: Scriver CR, Beaudet AL, Sly WS, Valle D (eds) The metabolic basis of inherited disease. McGraw-Hill, New York, pp 1045–1075
3. Daddona PE, Shewach DS, Kelley WN, Argos P, Markham AF, Orkin SH (1984) Human adenosine deaminase cDNA and complete primary amino acid sequence. J Biol Chem 259: 12101–12106
4. Orkin SH, Dadonna PE, Shewach DS, Markham AF, Bruns GA, Goff SC, Kelley W (1983) Molecular cloning of human adenosine deaminase gene sequences. J Biol Chem 258: 12753–12756
5. Valerio D, Duyvesteyn MGC, Meera Kahn P, van Kessel AG, de Waard A, van der Eb A (1983) Isolation of cDNA clones for human adenosine deaminase. Gene 25: 231–240
6. Wiginton DA, Adrian GS, Friedman D, Suttle DP, Hutton JJ (1983) Cloning of cDNA sequences of human adenosine deaminase. Proc Natl Acad Sci USA 80: 7481–7485
7. Wiginton DA, Adrian GS, Hutton JJ (1984) Sequence of human adenosine deaminase cDNA including the coding region and a small intron. Nucl Acid Res 12: 2439–2446
8. Williams SR, Goddard JM, Martin DWJ (1984) Human purine nucleoside phosphorylase cDNA sequence and genomic clone characterization. Nucl Acids Res 12: 5779–5787
9. Goddard JM, Caput D, Williams SR, Martin DWJ (1983) Cloning of human purine nucleoside phosphorylase cDNA sequences by complementation in Escherichia coli. Proc Natl Acad Sci USA 80: 4281–4285
10. Wiginton DA, Kaplan DJ, States JC; Akeson AL, Perme CM, Bilyk IJ, Vaughn AJ, Lattier DL, Hutton JJ (1986) Complete sequence and structure of the gene for human adenosine deaminase. Biochem 25: 8234–8244

11. Ealick SE, Rule SA, Carter DC, Greenhough TJ, Babu YS, Cook WJ, Habash J, Helliwell JR, Stoeckler JD, Parks REJ, Chen S-F, Bugg CE (1990) Three dimensional structure of human erythrocytic purine nucleoside phosphorylase at 3.2 Å resolution. J Biol Chem 265: 1812–1820

12. Wilson DK, Rudolph FB, Quiocho FA (1991) Atomic structure of adenosine deaminase complexed with a transition-state analog: understanding catalysis and immunodeficiency mutations. Science 252: 1278–1284

13. Jahnwar SC, Berkvens TM, Breukel C, van Ormondt H, van der Eb AJ, Kahn PM (1989) Localization of human adenosine deaminase (ADA) gene sequences to the q12–q13.11 region of chromosome 20 by in situ hybridization. Cytogenet Cell Genet 50: 168–171

14. Chang ZY, Nygaard P, Chinault AC, Kellems RE (1991) Deduced amino acid sequence of Escherichia coli adenosine deaminase reveals evolutionarily conserved amino acid residues: implications for catalytic function. Biochem 30: 2273–2280

15. Markert ML, Hutton JJ, Wiginton DA, States JC, Kaufman RE (1988) Adenosine deaminase (ADA) deficiency due to deletion of the ADA gene promoter and first exon by homologous recombination between two Alu elements. J Clin Invest 81: 1323–1327

16. Berkvens TM, van Ormondt H, Gerritsen EJ, Meera Khan P, van der Eb AJ (1990) Identical 3250–bp deletion between two AluI repeats in the ADA genes of unrelated ADA-SCID patients. Structural and functional analysis of the murine adenosine deaminase gene. Genomics 7: 476–485

17. Akeson AL, Wiginton DA, Dusing MR, States JC, Hutton JJ (1988) Mutant human adenosine deaminase alleles and their expression by transfection into fibroblasts. J Biol Chem 263: 16291–16296

18. Akeson AL, Wiginton DA, States JC, Perme CM, Dusing MR, Hutton JJ (1987) Mutations in the human adenosine deaminase gene that affect protein structure and RNA splicing. Proc Natl Acad Sci USA 84: 5947–5951

19. Bonthron DT, Markham AF, Ginsberg D, Orkin SH (1985) Identification of a point mutation in the adenosine deaminase gene responsible for immunodeficieny. J Clin Invest 76: 894–897

20. Hirschhorn R, Chakravarti V, Puck J, Douglas SD (1991) Homozygosity for a newly identified missense mutation in a patient with very severe combined immunodeficiency due to adenosine deaminase deficiency (ADA-SCID). Am J Hum Genet 49: 878–885

21. Markert ML, Norby SC, Ward FE (1989) A high proportion of ADA point mutations associated with a specific alanine-to-valine substitution. Am J Hum Genet 45: 345–361

22. Valerio D, Dekker BMM, Duyvesteyn MGC, van der Voorn L, Berkvens TM, van Ormondt H, van der Eb AJ (1986) One adenosine deaminase allele in a patient with severe combined immunodeficiency contains a point mutation abolishing enzyme activity. Embo J 5: 113–119

23. Hirschhorn R, Tzall S, Ellenbogen A, Orkin SH (1989) Identification of a point mutation resulting in a heat-labile adenosine deaminase (ADA) in two unrelated children with partial ADA deficiency. J Clin Invest 83: 497–501

24. Hirschhorn R, Tzall S, Ellenbogen A (1990) Hot spot mutations in adenosine deaminase deficiency. Proc Natl Acad Sci USA 87: 6171–6175

25. Arredondo-Vega FX, Kurtzberg J, Chaffee S, Santisteban I, Reisner E, Povey MS, Hershfield MS (1990) Paradoxical expression of adenosine deaminase in T cells cultured from a patient with adenosine deaminase deficiency and combined immunodeficiency. J Clin Invest 86: 444–452

26. Polmar SH, Stern RC, Schwartz AL, Wetzler EM, Chase PA, Hirschhorn R (1976) Enzyme replacement therapy for adenosine deaminase deficiency and severe combined immunodeficiency. N Engl J Med 295: 1337–1343

27. Polmar SH (1980) Metabolic aspects of immunodeficiency disease. Semin Hematol 17: 30–43

28. Hirschhorn R (1990) Adenosine deaminase deficiency. In: Rosen FS, Seligmann M (eds) Immunodeficiency reviews. Harwood Academic, New York, pp 175–198

29. Fischer A, Landais P, Friedrich W, Morgan G, Gerritsen B, Fasth A, Porta F, Griscelli C, Goldman SF, Levinsky R, Vossen J (1990) European experience of bone marrow transplantation for severe combined immunodeficiency. Lancet 336: 850–854

30. O'Reilly RJ, Keever CA, Small TN, Brochstein J (1989) The use of HLA-non-identical T-cell-depleted marrow for transplant for correction of severe combined immunodeficiency disease. Immunodeficiency Rev 1: 273–309

31. Hershfield MS (1991) Enzyme replacement therapy for inherited metabolic diseases. In: Friedmann T (ed) Therapy for genetic disease. Oxford University Press, Oxford, pp 76–94

32. Abuchowski A, McCoy JR; Palczuk NC, van Es T, Davis FF (1977) Effect of attachment of polyethylene glycol on immunogenicity and circulating life of bovine liver catalase. J Biol Chem 252: 3582–2586

33. Abuchowski A, van Es T, Palczuk NC, Davis FF (1977) Alteration of immunological properties of bovine serum albumin by covalent attachment of polyethylene glycol. J Biol Chem 252: 3578–3581

34. Davis S, Abuchowski A, Park YK, Davis FF (1981) Alteration of the circulating life and antigenic properties of bovine adenosine deaminase in mice by attachment of polyethylene glycol. Clin Exp Immunol 46: 649–652

35. Hershfield MS, Buckley RH, Greenberg ML, Melton AL, Schiff R, Hatem C, Kurtzberg J, Markert ML, Kobayashi RH, Kobayashi AL, Abuchowski A (1987) Treatment of adenosine deaminase deficiency with polyethylene glycol-modified adenosine deaminase. N Engl J Med 316: 589–596

36. Bory C, Boulieu R, Souillet G, Chantin C, Rolland MO, Mathieu M, Hershfield MS (1990) Comparison of red cell transfusion and polyethylene glycol-modified adenosine deaminase therapy in an adenosine deaminase-deficient child. Ped Res 28: 127–130

37. Levy Y, Hershfield MS, Fernandez-Mejia C, Polmar SH, Scudiery D, Berger M, Sorensen RU (1988) Adenosine deaminase deficiency with late onset of recurrent infections: response to treatment with polyethylene glycol-modified adenosine deaminase (PEG-ADA). J Pediatr 113: 312–317

38. Hershfield MS, Chaffee S (1991) PEG-enzyme replacement therapy for adenosine deficiency. In: Desnick RJ (ed) Treatment of genetic disease. Churchill Livingstone, New York, pp 169–182

# IV   The Purine Nucleotide Cycle

# IV A  Myoadenylate (Muscle AMP) Deaminase Deficiency

## 1  Clinical Aspects and Biochemical Basis of AMP Deaminase Deficiency: A Clinician's Point of View

N. Zöllner, D. R. Wagner, and M. Gross

In 1978 adenosine monophosphate (AMP) deaminase deficiency in skeletal muscle ("myoadenylate deaminase deficiency") was described as a new entity by Fishbein et al. Their publication, "Myoadenylate deaminase deficiency: a new disease of muscle," reported five patients with muscular weakness or cramping after exercise. Muscle biopsies of all five patients showed less than 5% of normal AMP deaminase activity, levels of other enzymes were normal. Routine histology of skeletal muscle was also normal. The symptoms of these patients were attributed to the lack of AMP deaminase activity.

During the first 5 years after the report by Fishbein et al.(1978), more than 100 cases of myoadenylate deaminase (MAD) deficiency were published in the USA and Europe. Obviously, the disease is not rare. Clinical features and the pathogenesis of symptoms are not yet understood.

## Clinical Aspects

### Case Report

The first patient seen by one of us (N.Z.) in Munich was the first case diagnosed clinically in Germany. This patient and the symptomatic effect of ribose have previously been reported (Zöllner et al. 1986, Gross et al. 1991). When seen in 1984, he was a 55-year-old accountant who complained of long-lasting, very painful, exercise induced stiffness of his muscles. Until 1982 the patient had been a real sportsman, swimming, cycling in hilly areas, and skiing downhill and cross country. After an interruption of these activities in 1982 due to hospitalization because of a bleeding duodenal ulcer, he felt muscle pain and stiffness after exercise. He thought that the interruption of his sporting activities might have caused this decrease in muscular performance. Therefore, he unsuccessfully tried to regain his former endurance by training, but symptoms got worse. When he was skiing downhill, the pain in the legs became so severe that he had to lay down in the snow at the end of a turn because it was too painful to stand. After resting for several minutes he could go for another turn, but the pain became worse. Finally, he had to give up skiing. He noted that pain and stiffness primarily involved the muscles exercised. In 1983 the patient had to a-bandon all sporting activities. He sought medical advice from a variety of physi-cians but no diagnosis was made. Repeatedly he was considered a malingerer.

To us, some of his self observations strongly supported the diagnosis of a dis-order in muscular energy metabolism: Any skeletal muscle became painful if exercised. For example, playing the accordion caused pain in the forearms. The pain occurred during intensive exercise, and it lasted longer than muscular pain caused by obliterating vascular disease. After heavy work the pain persisted for several days, although decreasing in intensity. A muscle biopsy was taken; it was reported normal. Histochemical and biochemical analysis revealed a lack of MAD activity.


## Discussion

In patients with MAD deficiency or other metabolic myopathies, the history is most important for the correct diagnosis. It matters whether pain on (not after!) exertion is related to the intensity of work, whether it is related to the muscles exercised, and whether training diminishes the tendency for pain or not. A few precise questions may save a lot of analyses.

MAD deficiency is the most common of all muscle enzyme defects known so far. It is found in about 2%–3% of all patients in large series of muscle biopsies (Sabina et al. 1989; Fishbein 1986).

The mean age at time of diagnosis is 32 years, with a range from 1 to 70 years. About one-fourth of all patients develop symptoms either in infancy or child-hood, as teenagers, as young adults or at the age of 40 years or older (Sabina et al. 1989).

The symptoms of MAD deficiency are not specific for this disease. They indi-cate a disorder in muscular energy metabolism. Unlike myophosphorylase defi-ciency, which can be diagnosed clinically by the "second wind" phenomenon, there are no clinical features of MAD deficiency that might help to differentiate this enzyme defect from most of the other metabolic disorders.

Physical and neurological examinations reveal no abnormalities in patients with MAD deficiency. The routine laboratory tests exhibit normal results except for a slightly elevated serum creatine phosphokinase level in about half of the patients. Technical examinations including electromyography do not help to establish the diagnosis, and electromyograms may show some unspecific minor changes. Histological examination including electron microscopy may reveal small alterations of the muscle that are not specific for MAD deficiency.

Patients with muscular symptoms can be screened for MAD deficiency with the ischemic exercise test of Munsat (1970). Since the reaction catalyzed by MAD is the main source of ammonia in working muscle, patients with MAD de-ficiency do not produce ammonia during exercise but lactate production is not altered. The diagnosis is usually confirmed by histochemical and biochemical examination of a muscle biopsy. In the last year the mutation $C^{34}$-T of the AMPD1 gene was found to be the genetic basis of all patients with MAD deficiency studied so far. Methods for the detection of this mutation were developed. They now enable us to diagnose patients with MAD deficiency without taking a muscle biopsy (Gross 1992).

Our first patient's symptoms were typical of a metabolic myopathy. Most patients suffer from myalgia during exercise, early fatigue, and postexercise symptoms including cramps and muscle pain. These complaints were reported by 49 of the first 56 patients (88%) reported in the literature (Sabina et al. 1989). These figures may not represent the full range of symptoms in MAD deficiency since there is a tendency to report mainly impressive symptoms.

Indeed, some patients with MAD deficiency report symptoms other than exercise-induced muscle pain, cramping, or early fatigue. Many of the patients seen in our hospital reported muscle pain at rest that is aggravated by exercise. Case reports have been published on MAD deficient patients with permanent muscular hypotonia, rhabdomyolysis, or delayed motor development (Ashwal and Peckham 1985; Fishbein 1986; Sabina et al. 1989). In many patients, MAD deficiency is combined with other muscular disorders (Goebel and Bardosi 1987). MAD deficiency was reported in patients suffering from muscular dystrophy, neuropathy, polymyositis, muscle atrophy, or connective tissue disease. The lack of MAD activity in muscle biopsies of these patients was considered a secondary phenomenon, particularly when enzymes other the MAD were also of lowered activity.

A correlation between MAD deficiency and myalgia or exercise-induced muscle symptoms was found in two studies. Kelemen et al. (1982) analyzed 302 muscle biopsies: 36 were taken because of myalgias, and three of these (8.3%) were MAD deficient. Joosten et al. (1984) analyzed 16 muscle biopsies from patients with exercise-induced muscle symptoms, and two (12.5%) were MAD deficient. However, other studies did not find such a high correlation. None of 35 patients with aches, cramps, and pain in the muscles had MAD deficiency in the study of Mercelis et al. (1987). Kelemen et al. (1982) reported family studies. They found some members of families of patients with MAD deficiency that reported myalgias. These family members had normal MAD activities in muscle biopsies.

Some investigators like Fishbein deny the need of therapy: "The prognosis in myoadenylate deaminase deficiency appears to be excellent, with no evidence for progressive debilitation or structural damage in the absence of other disease, and therapy should be aimed at reassurance, and the avoidance of hazardous and ineffective long-term drug treatment" (Fishbein 1984, p.80). Our first patient could not agree with this statement. His life was heavily hampered by his disease, and he was looking for any possible therapy.

A rational therapy of enzyme defects must be based on the biochemical consequences of the defect and the pathogenesis of its symptoms. The first study of MAD deficiency detected a loss of most of the adenine nucleotides in working skeletal muscle (Sabina et al. 1980). Based on these findings, ribose was given to patients with MAD deficiency by two investigators in doses of 2 g daily (Patten 1982; Lecky 1983). The results of these first trials were negative, unconvincing at least. Since 2 g daily (8 kcal, 0.4% of the energy requirement) seemed to be a small amount of carbohydrates in comparison with the muscle mass, we started a regimen with up to 100 g ribose daily. In a double-blind trial 50 g daily of ribose or xylitol prevented or relieved the muscle pain; glucose and sorbitol were ineffective. When the patient took ribose orally in a dose of 15–20 g/h during exercise the muscle symptoms did not occur (for practical reasons, he dissolved ribose in tea or water and took a big sip every 5 to 10 min). Today, after years of therapy, he is much better: he can carry on his normal daily life, including walks or playing the accordion, without pain, although he does not take ribose every day.

The efficacy of ribose in MAD deficiency cannot easily be explained by the pathways known (Gross and Gresser 1993). Therefore, the effect of ribose may indicate that the contemporary hypotheses on energy metabolism in working muscle are incomplete or inaccurate.


## Biochemical Basis

AMP deaminase is one of three enzymes in the purine nucleotide cycle (Fig 1). This cycle also includes the reactions of adenylosuccinate synthetase and the adenylosuccinate lyase. The "purine nucleotide cycle" was first postulated by Lowenstein and Tornheim in 1971. There is evidence that it not only operates in skeletal muscle but also in other tissues like brain (Schultz and Lowenstein 1976). The symptoms in patients with MAD deficiency were explained as consequences of an interruption of the purine nucleotide cycle.

In an early study of a MAD-deficient patient (36-year-old female), muscle biopsies were taken prior to and after 15 min of step-up exercise on a platform at 17–19 steps per minute. After exercise, a large drop in adenine nucleotide concentration in myocytes (Sabina et al. 1980; Swain et al. 1983) was found.

Another study with bicycle ergometer exercise until exhaustion and/or muscle discomfort included four patients with MAD deficiency (3 female, 1 male), eight patients with exercise-induced muscle symptoms but normal MAD activity, and 11 healthy control subjects. Needle biopsies were taken from the

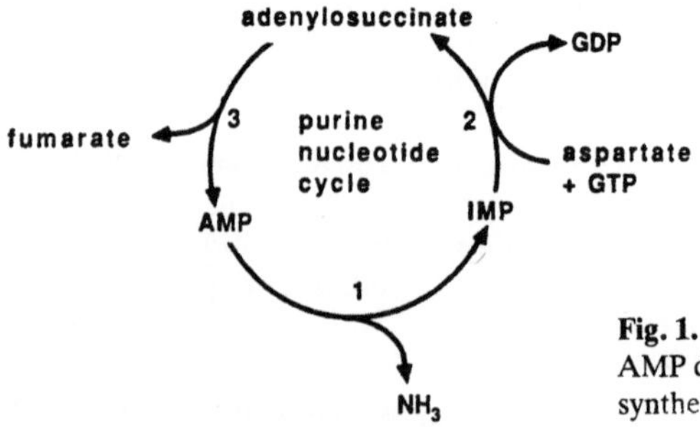

**Fig. 1.** The purine nucleotide cycle: 1. AMP deaminase; 2. Adenylosuccinate synthetase; 3. Adenylosuccinate lyase

m. vastus lateralis before and immediately after exercise (Sabina et al. 1984). The MAD-deficient patients could only perform 28% of the ergometer exercise done by patients with normal MAD activity. At the end of exercise the muscle creatine phosphate had decreased by 69% in the MAD-deficient patients and by 52% in the patients with normal MAD activity. Muscle ATP decreased by 6% in the MAD-deficient patients and by 34% in the patients with normal MAD activity. The decrease in ATP and creatine phosphate in comparison to the work performed was five times higher in the MAD-deficient patients than in the other patients. Muscle adenosine increased 16 fold in the MAD-deficient patients and only 2 fold in the patients with normal MAD activity. The muscle IMP concentration increased from 0.07 to 4.1 nmol/µmol creatine and from 0.38 to 51.6 nmol/µmol creatine, respectively. The muscle ATP content decreased by more than 20% in one of four MAD-deficient patients and in seven of eight patients without MAD deficiency. In the MAD-deficient patient, the increases in ADP, AMP, adenosine, inosine, and hypoxanthine corresponded to less than half of the drop in ATP, indicating a loss of adenine nucleotides from the patient's myocytes. In the control patients, the drop in ATP was equivalent to the increase in ADP, AMP, adenosine, inosine, and hypoxanthine.

Muscle biopsies were taken from five patients with MAD deficiency (4 men, 1 women) and ten healthy control subjects before and after exercise from the m. quadriceps femoris (Sinkeler et al. 1987). The subjects performed an isometric contraction with a knee extensor at 50% of the maximal voluntary contraction force until exhaustion. During contraction the blood flow was interrupted by inflating a cuff around the upper leg to 240 mmHg. There was no difference in maximal contraction force or endurance time between patients and control subjects. Skeletal muscle ATP and the sum of ATP, ADP, and AMP was higher in the MAD-deficient patients before and after exercise. The exercise-induced decrease of muscle ATP was less in the patients than in the control subjects; the decrease in creatine phosphate and the increase in muscle lactate were not different. The energy expenditure per unit of performance calculated from the decrease in ATP and creatine phosphate and the increase in lactate were not different. The energy expenditure per unit of performance calculated from the decrease in ATP and creatine phosphate and the increase in lactate was not different between the MAD-deficient patients and the control subjects.

The contradictory findings of these two studies with different types of exercise may be due to differences in the energy metabolism of isometric contraction on the one side, and exercise on the other. However, differences in the clinical severity of the two patient groups cannot be excluded.

It is known that the de novo synthesis of adenine nucleotides is limited by the availability of phosphoribosylpyrophosphate. The pool of phosphoribosylpyrophosphate and thereby the synthesis of adenine nucleotides can be increased in myocardial cells by giving ribose (Zimmer and Ibel 1984). The effect of ribose on adenine nucleotide synthesis in myocytes has not yet been studied. Our experiences with ribose and xylitol may provide important information concerning the metabolic changes in MAD deficiency.

# References


Ashwal S, Peckham N (1985) Myoadenylate deaminase deficiency in children. Pediat Neurol 1: 185–191

Fishbein WN (1984) Human myoadenylate deaminase deficiency. Adv Exp Med Biol 165A: 77–84

Fishbein WN (1986) Myoadenylate deaminase deficiency. In: Engel AG, Banker BQ (eds) Myology. McGraw-Hill, New York, pp 1745–1762

Fishbein WN, Armbrustmacher VW, Griffin JL (1978) Myoadenylate deaminase deficiency: a new disease of muscle. Science 200: 545–548

Goebel HH, Bardosi A (1987) Myoadenylate deaminase deficiency. Klin Wochenschr 65: 1023–1033

Gross M, Kormann B, Zöllner N (1991) Ribose administration during exercise: Effects on substrates and products of energy metabolism in healthy subjects and a patient with myoadenylate deaminase deficiency. Klin Wochenschr 69: 151–155

Gross M, Gresser U (1993) Ergometer exercise in myoadenylate deaminase deficient patients. Clin Investig 71: 461–465

Gross M (1992) Genetic basis of myoadenylate deaminase deficiency in man. (this volume)

Joosten E, van Bennekom C, Oerlemans F, de Bruyn C, Oei T, Trijbels J (1984) Myoadenylate deaminase deficiency: an enzyme defect in search of a disease. Adv Exp Med Biol 165A: 85–89

Kelemen J, Rice DR, Bradley WG, Munsat TL, DiMauro S, Hogan EL (1982) Familial myoadenylate deaminase deficiency and exertional myalgia. Neurology 32: 857–863

Lecky BRF (1983) Failure of D-ribose in myoadenylate deaminase deficiency. Lancet 1: 193

Lowenstein J, Tornheim K (1971) Ammonia production in muscle: the purine nucleotide cycle. Science 171: 397–400

Mercelis R, Martin JJ, de Barsy T, van de Berghe G (1987) Myoadenylate deaminase deficiency: absence of correlation with exercise intolerance in 452 muscle biopsies. J Neurol 234: 385–389

Munsat TL (1970) A standardized forearm ischemic exercise test. Neurology 20: 1171–1178

Patten BM (1982) Beneficial effect of D-ribose in patient with myoadenylate deaminase deficiency. Lancet 1: 107

Sabina RL, Swain JL, Patten BM, Ashizawa T, O'Brien WE, Holmes EW (1980) Disruption of the purine nucleotide cycle. A potential explanation for muscle dysfunction in myoadenylate deaminase deficiency. J Clin Invest 66: 1419–1423

Sabina RL, Swain JL, Olanow CW, Bradley WG, Fishbein WN, DiMauro S, Holmes EW (1984) Myoadenylate deaminase deficiency. Functional and metabolic abnormalities associated with disruption of the purine nucleotide cycle. J Clin Invest 73: 720–730

Sabina RL, Swain JL, Holmes EW (1989) Myoadenylate deaminase deficiency. In: Scriver CR, Beaudet AL, Sly WS, Valle D (eds) The metabolic basis of inherited disease, 6th edn. McGraw-Hill, New York, pp 1077–1184

Schultz V, Lowenstein JM (1976) Purine nucleotide cycle. Evidence for the occurrence of the cycle in brain. J Biol Chem 251: 485–492

Sinkeler SPT, Binkhorst RA, Joosten EMG, Wevers RA, Coerwinkel MM, Oei TL (1987) AMP deaminase deficiency: study of the human skeletal muscle purine metabolism during ischaemic isometric exercise. Clin Sci 72: 475–482

Swain JL, Sabina RL, Holmes EW (1983) Myoadenylate deaminase deficiency. In: Stanbury JB, Wyngaarden JB, Fredrickson DS, Goldstein JL, Brown MS (eds) The metabolic basis of inherited disease, 5th edn. McGraw-Hill, New York, pp 1184–1191

Zimmer HG, Ibel H (1984) Ribose accelerates the repletion of the ATP pool during recovery from reversible ischemia of the rat myocardium. J Mol Cell Cardiol 16: 863–866

Zöllner N, Reiter S, Gross M, Pongratz D, Reimers CD, Gerbitz K, Paetzke I, Deufel T, Hübner G (1986) Myoadenylate deaminase deficiency: Successful symptomatic therapy by high dose oral administration of ribose. Klin Wochenschr 64: 1281–1290

# 2 The AMP Deaminase Multigene Family in Rats and Humans

D. K. Mahnke-Zizelman, M. T. Bausch-Jurken, and R. L. Sabina

## Introduction

Adenosine monophosphate (AMP) deaminase is a purine nucleotide interconverting enzymatic activity critical to energy metabolism, e.g., its role in the proposed functions of the purine nucleotide cycle (Lowenstein 1972). In higher eukaryotes, multiple isoforms exhibiting tissue-specific and developmental patterns of expression have been described. In addition to exhibiting diverse biochemical and immunological properties, limited cellular biology studies suggest that intraspecies variants of AMP deaminase occupy different intracellular addresses. Recent molecular studies have demonstrated AMP deaminase isoforms to be manifest through the expression of a multigene family, some members of which produce primary transcripts subject to alternative splicing. Clinically, inherited and acquired deficiencies of skeletal muscle AMP deaminase (Sabina et al. 1989b) and an inherited deficiency in erythrocytes (Ogasawara et al. 1984b) have been described. Our laboratory is utilizing cellular and molecular approaches to gain insight into the functional significance of AMP deaminase isoform diversity.

## AMP Deaminase Isoforms in Rats and Humans

Before discussing the molecular biology of AMP deaminase expression, it is useful to review available protein literature. Appropriately, this is most extensive in rats and humans. Biochemical and immunological studies have detailed three isoforms of AMP deaminase in rats and four in humans. The combined data indicate similarities among the AMP deaminase isoforms expressed in these two mammalian species.

### Rats

Three AMP deaminase variants have been described in rat tissues and are termed isoforms A, B, and C (Ogasawara et al. 1975a). Isoform A (myoadenylate deaminase) is the main activity found in adult skeletal muscle. Isoform B predominates in liver, kidney, and testes. Isoform C is isolable from cardiac muscle. Table 1 lists reported tissue and cell distributions for the various rat activities as determined by a combination of chromatographic, electrophoretic, and immunological techniques. Although isoform A is known to be expressed only in postnatal skeletal muscle and well-differentiated skeletal myocytes in culture, isoforms B and C exhibit wide tissue distributions. Purified native isoforms A (Coffee and Kofke 1975; Ogasawara et al. 1977; Stankiewicz et al. 1979) and B (Ogasawara et al. 1977) are tetramers composed of identical subunits. Reported subunit molecular weights for isoform A range from 60K to 80K (Coffee and Kofke 1975; Ogasawara et al. 1977; Wheeler and Lowenstein 1979; Stankiewicz et al. 1979; Marquetant et al. 1989; Tovmasian et al. 1990). Purifications performed in the presence and absence of proteolytic inhibitors strongly suggest that proteolysis is a major determinant in generating these variable subunit molecular weights (Marquetant et al. 1989). Isoform B, purified from adult liver, has a reported subunit molecular weight of 85K (Ogasawara et al. 1977). Isoform C has been purified from adult heart (Ogasawara et al. 1975a; Kaletha and Skladanowski 1979) but subunit molecular weights were not reported.

Chromatographic and immunological data indicate hybrid activities in those tissues and cells exhibiting more than one AMP deaminase isoform (Ogasawara et al. 1975a and 1975b; Sabina et al. 1989b). For example, phosphocellulose chromatography can be utilized to separate five peaks of AMP deaminase activity in rat brain (Ogasawara et al. 1975a and 1975b). Immunological analyses indicate that three of these peaks are isoform B/C hybrids whereas the other two peaks are reactive only with a single isoform-specific antiserum (Ogasawara et al. 1975b).

All AMP deaminase activities in rats are allosterically regulated by alkali metal ions and nucleoside di- or triphosphates (Ogasawara et al. 1979). However, each isoform displays unique kinetic and regulatory characteristics including different substrate affinities and specificities (Ogasawara et al. 1975a) and responses to alkali metal ions and nucleotide affectors, i.e., ATP, GTP, and adenylate energy charge (Ogasawara et al. 1979). Covalent modification of AMP deaminase may also play a role in its allosteric regulation. For example, protein kinase C-induced phosphorylation of isoform A, in vitro, results in activation of the enzyme (Tovmasian et al. 1990).

AMP deaminase isoforms also exhibit separate patterns of developmental expression in rat tissues and cells. Skeletal myogenesis, in vivo and in vitro, is characterized by the temporal appearance of isoforms B → C → A (Marquetant et al. 1987; Sabina et al. 1989a), with the latter two being coexpressed in adult myofibers (Thompson et al. 1992). Relative to skeletal myogenesis, rat AMP deaminase isoforms have been referred to as embryonic (isoform B), perinatal (isoform C), and adult (isoform A) (Marquetant et al. 1987). The relative proportions of isoforms A and C vary in adult skeletal muscle depending on fiber type (Thompson et al. 1992). Almost exclusively, glycolytic muscle expresses isoform A, whereas isoform C constitutes up to one third of the total AMP deaminase activity in oxidative muscle. Although not as well documented, brain, heart, kidney, liver, and lung exhibit developmental changes in the relative proportions of isoforms B and C in newborn versus adult tissues (Ogasawara et al. 1975b and 1978a).

The cellular biology of rat AMP deaminase is beginning to emerge. Histo-

**Table 1.** Distribution of rat and human AMP deaminase isoforms

| Tissue/Cell Type | Rats | Humans |
|---|---|---|
| Skeletal Muscle | | |
|   Embryonic | B | ? |
|   Perinatal | B, C[a] | ? |
|   Adult | A, C | M, E1 |
| Skeletal Myocytes | | |
|   Myoblasts | B (L6) | ? |
|   Myotubes | A, B, C (L6) | ? |
| Brain | | |
|   Newborn | B, C[b] | ? |
|   Adult | B, C | L, E1[c] |
| Heart | | |
|   Newborn | C, B | ? |
|   Adult | C, B | E1, L |
| Kidney | | |
|   Newborn | B, C | ? |
|   Adult | B, C | L, E1 |
| Liver | | |
|   Newborn | B, C | ? |
|   Adult | B, C | L, E1 |
| Lung | | |
|   Newborn | B, C | ? |
|   Adult | B, C | L, E1 |
| Spleen (Adult) | B, C | E1, L |
| Testis (Adult) | B, C | ? |
| Erythrocyte | C | E1 |
| Granulocyte | ? | E1 |
| Platelet | ? | L |
| Mononuclear cell | ? | L |
| T lymphoblast | ? | L |
| B lymphoblast | ? | L, E1 |

[a] In those cases in which more than one isoform is detected the predominant activity is listed first.

[b] When newborn and adult tissues are compared there is generally a developmental increase in the percentage of the dominant activity and a corresponding decrease in the lesser activity.

[c] Isoform E2 is generally detected as a minor component of isoform E1.

chemical analyses of adult rat skeletal muscle demonstrate intermyofibrillar and A-band staining (Meyer et al. 1980; Thompson et al. 1992), the latter consistent with the reported tight binding of isoform A to myosin (Shiraki et al. 1979; Marquetant et al. 1989). Whether this interaction with myosin is an isoform-specific property is controversial (see also Ogasawara et al. 1978b). Histochemical staining of AMP deaminase activity is also observed in other elements of skeletal muscle, particularly in intrafusal skeletal muscle fibers, the Schwann sheath of nerve fibers, in the intima and media of arteries, and in the capillary wall (Gilloteaux and Meyer 1983). Immunofluorescent analysis of adult rat skeletal muscle (Thompson et al. 1992) reveals that isoform A is responsible for the myofibrillar-associated activity. Subsarcolemmal staining is also observed with anti-A serum. Anti-C serum primarily reacts with vascular elements, such as arterial and venous walls, capillary endothelium, and erythrocytes. However, significant subsarcolemmal staining is also evident. Anti-B serum is relatively less reactive, but quite specifically associated with connective tissues surrounding neural elements and the muscle spindle capsule.

## Humans

Four AMP deaminase activities can be distinguished in human tissues and cells. These include isoform M(uscle), L(iver), E1 and E2 (erythrocyte) (Ogasawara et al. 1982). Human isoforms of AMP deaminase display patterns of expression similar to the rat activities (Table 1), data that indicate the following cross-species relationships: isoform M (rat A), isoform L (rat B), and isoforms E1 and E2 (rat C). This contention is generally consistent with reported immunological properties of the various rat and human activities. Anti-A and anti-B sera, each specific for the appropriate rat isoform (Ogasawara et al. 1975), will also recognize human isoforms M and L, respectively (Ogasawara et al. 1982). However, anti-C serum does not cross-react with any other rat (Ogasawara et al. 1975a) or human (Ogasawara et al. 1982) AMP deaminase isoform.

Human AMP deaminase isozymes can be distinguished by phosphocellulose chromatography and zone electrophoresis (Ogasawara et al. 1982). Kinetic and immunological analyses of isoforms M and L demonstrate that they differ from each other and from the erythrocyte variants (Ogasawara et al. 1982). Isoform M has been purified to apparent homogeneity from adult skeletal muscle and has a reported subunit molecular weight of 71K-72K (Stankiewicz 1981; Ogasawara et al. 1982). Isoform L has been purified to apparent homogeneity from autopsied adult liver (Ogasawara et al. 1982) and outdated platelets (Ashby and Holmsen 1981), with reported subunit molecular weights of 68K and 83K–85K, respectively. Isoform E1 has been purified to apparent homogeneity from outdated erythrocytes and has a reported subunit molecular weight of 80K (Yun and Suelter 1978; Ogasawara et al. 1982). Purified human AMP deaminase isoforms appear to have tetrameric structures (Ogasawara et al. 1982). Isoform E2, partially purified from outdated erythrocytes, can be distinguished chromatographically but not kinetically or immunologically from isoform E1 (Ogasawara et al. 1982). Their coexpression in human tissues and cells (Ogasawara et al. 1982 and 1984a) and combined absence in an inherited erythrocyte deficiency (Ogasawara et al. 1984b) have led to the proposal that isoforms E1 and E2 are products of the same gene (Ogasawara et al. 1984b).

Developmental expression of human AMP deaminase isoforms is not as well characterized. Total AMP deaminase activity reportedly increases during human skeletal myogenesis in vivo (Kaletha et al. 1987) and in vitro (Benders et al. 1991). Chromatographic analyses of 11 and 16 week old fetal, and adult human skeletal muscle extracts demonstrate developmental changes in the relative amounts of two well-separable peaks of AMP deaminase activity (Kaletha et al. 1987). Kinetic analyses of these partially purified activities indicate a shift from a "fetal" isoform to an "adult" skeletal muscle isoform (Kaletha and Nowak 1988).

The cellular biology of human AMP deaminase is also not as well documented. Similar to rats and other higher eukaryotic species, histochemical staining of skeletal muscle indicates a concentration of AMP deaminase activity in the myocyte (Fishbein et al. 1980). However, localization of isoform M at the level of the myofibril has never been reported. In vitro studies suggest that AMP deaminase (presumably isoform E1 and/or E2) binds preferentially and specifically to the cytoplasmic surface of the erythrocyte membrane (Rao et al. 1968; Pipoly et al. 1979). No information is available regarding the cellular biology of isoform L.

## AMP Deaminase Genes in Rats and Humans

Molecular biological studies of AMP deaminase have generated new and potentially exciting information regarding isoform diversity. To date, three AMP deaminase genes have been identified in rats and humans. Although comprehensive analyses have yet to be reported, it is likely that the AMP deaminase multigene family is regulated by tissue-specific and developmental factors. Alternative splicing events in at least two of these genes further imply that AMP deaminase expression is likely to be more complex than indicated by the protein literature.

### Rats

Isoform A (myoadenylate deaminase) cDNA was the first AMP deaminase-specific nucleic acid reagent to be isolated and characterized (Sabina et al. 1987). The entire open reading frame of myoadenylate deaminase cDNA predicts a polypeptide with a subunit molecular weight of 87K. Northern blot analyses demonstrate skeletal muscle specificity and developmental expression of myoadenylate deaminase transcript. Southern blot analyses indicate that the myoadenylate deaminase gene is highly conserved across eukaryotic species. Subsequently, myoadenylate deaminase cDNA was employed as the probe to isolate the gene, designated *AMPD1* (Sabina et al. 1990). This gene is approximately 21 kb and is comprised of 16 exons. Sequences within 200 bp 5' of the transcription start site contain information necessary to direct muscle-specific expression of AMP deaminase. Analyzed in greater detail, the region between $-100$ and $-79$ was found to contain a myocyte-specific enhancer binding factor (MEF-2)-like binding site (Cserjesi and Olson 1991) which exhibits a DNase I footprint with myocyte nuclear extract (Morisaki et al. 1991a).

The rat *AMPD1* gene produces two mature 2.5 kb transcripts, one that retains miniexon 2 and one from which exon 2 is removed. Developmental and tissue-specific patterns of this alternative splicing event are observed (Sabina et al. 1989a; Mineo et al. 1990) and attributed to factors involving both exon recognition and nucleocytoplasmic partitioning (Mineo and Holmes 1991). The functional significance of alternative splicing of the rat *AMPD1* primary transcript is unknown. The predicted N-terminal domain of isoform A is not conserved when compared to the predicted yeast protein (Meyer et al. 1989; Sabina et al. 1990). Furthermore, deletion of 215 bp from the 5' end of isoform A cDNA, which includes exon 2 sequences, does not eliminate catalytic activity of this truncated cDNA when placed in a prokaryotic expression vector (Mineo et al. 1990).

Employing 3' end sequences of isoform A cDNA to construct an antisense RNA probe, a third rat AMP deaminase transcript can be detected in embryonic hindlimb and proliferating skeletal myoblasts in culture by RNase protection analysis (Sabina et al. 1989a). Concomitant expression of this transcript with anti-B reactive AMP deaminase activity in several rat tissues and cells (Sabina et al. 1989a; Morisaki et al. 1990) implies specificity to isoform B. By mapping the protected portion of isoform A cRNA in these experiments, oligonucleotides were designed exhibiting high nucleotide similarity with isoform B transcript. Utilizing these oligonucleotides as probes, a cDNA clone was isolated from a rat brain library and partially characterized (Morisaki et al. 1990). Nucleotide sequence analyses demonstrate that the recombinant insert represents partial isoform B primary transcript. Included are 264 bp of coding sequence exhibiting similarity (69%) to isoform A cDNA. Southern blot analyses of rat genomic DNA generate independent banding patterns with isoform A and B cDNA probes. These results provide evidence for a second AMP deaminase gene in the rat, termed *AMPD2*. Additional Southern blot analyses of genomic DNA isolated from coformycin-resistant skeletal myocytes indicate that *AMPD1* and *AMPD2* are closely linked in the rat genome. Northern blot analysis of polyA+ RNA isolated from rat brain reveals that the *AMPD2* gene produces a 3.4 kb transcript.

Recently, a unique cDNA clone has been isolated from an adult rat soleus muscle library which is similar to isoform A and B cDNAs (D'Cuhna J, Mahnke-Zizelman DK, Sabina RL, unpublished work). Limited RNase protection analyses suggest this cDNA represents partial isoform C transcript. Southern blot analyses indicate a third AMP deaminase gene in the rat, tentatively termed *AMPD3*.

## Humans

Isoforms M (myoadenylate deaminase) cDNA is isolable from adult skeletal muscle libraries utilizing rat isoform A cDNA as the probe (Sabina et al. 1992). Alignment of isoform A and M cDNA sequences demonstrates identically sized open reading frames exhibiting 87% nucleotide similarity and 93% predicted amino acid homology. Isoform M cDNA was employed as the probe to isolate the human *AMPD1* gene (Sabina et al. 1990). Similar to the gene in the rat, human *AMPD1* is approximately 23 kb and is composed of 16 exons. Comparison of upstream nucleotide sequence indicates four conserved domains in rat and human *AMPD1* (Sabina et al. 1990), including one which contains an identical muscle-specific enhancer motif (Morisaki et al. 1991a; Cserjesi and Olson 1991). Unlike that documented for the rat *AMPD1* primary transcript, RNase protection analyses of adult human skeletal muscle RNA do not detect an alternative splicing event involving miniexon 2 (Mineo et al. 1990). Therefore, northern blot analyses are taken to indicate that the human *AMPD1* gene produces only a single 2.5 kb transcript (Sabina et al. 1991). In situ hybridization and Southern blot analysis of human-mouse somatic cell hybrids localizes the *AMPD1* gene to chromosome 1 in the region p13-p21 (Sabina et al. 1990), results that are consistent with an autosomal mode of inheritance for the skeletal myopathy. Accordingly, a single point mutation in miniexon 2 has been recently identified in 12 unrelated myoadenylate deaminase deficient patients (Morisaki et al. 1991b).

Utilizing rat *AMPD2* cDNA as the initial probe, a near full-lenght cDNA has been recently isolated from human T lymphoblast and placental libraries (Bausch-Jurken MT, Mahnke-Zizelman DK, Morisaki T, Sabina RL, unpublished work). The open reading frame contained within this cDNA produces anti-B (isoform L-specific) reactive AMP deaminase activity when placed in a prokaryotic expression vector. Subsequently, the human *AMPD2* gene has been isolated, sequenced, and localized (Bausch-Jurken MT, Mahnke-Zizelman DK, Eddy R, Shows TB, Sabina RL, unpublished work).

Additional cloning efforts, also employing rat *AMPD2* cDNA as the initial probe, have resulted in the isolation of a third human AMP deaminase cDNA from the same T lymphoblast library (Mahnke-Zizelman DK, Sabina RL, unpublished work). Sequence analysis of additional recombinants isolated from keratinocyte and vascular endothelial libraries documents heterogeneous 5' ends. Partial characterization of 5' end human *AMPD3* genomic clones identifies at least four exons. RNase protection analyses demonstrate alternative splicing of at least three 5' terminal exons in the *AMPD3* primary transcript. Combined, these results may explain the molecular basis of isoform E1 and E2 expression.

## Conclusions and Future Directions

Availability of near full-lenght cDNA clones for all known human AMP deaminase isoforms provides information that may be utilized to predict structure/ function relationships. Alignments of primary amino acid sequences predicted from *AMPD1*, *AMPD2*, and *AMPD3* cDNA nucleotide sequences demonstrate conservation only in the C-terminal two thirds of the polypeptides. Interestingly, identical residue in this portion of the AMP deaminases are highly homologous with those in the proposed active site of adenosine deaminase (Chang et al. 1991; Wilson et al. 1991). Moreover, AMP deaminase cDNAs from which 5' terminal sequences have been deleted are catalytically active when placed in a prokaryotic expression vector (Mineo et al. 1990). Combined, these data strongly suggest that the catalytic site of AMP deaminase is composed of residues located in the conserved C-terminal domain of the polypeptide. Verification of this hypothesis must await X-ray crystallographic studies of purified AMP deaminase and site-directed mutagenesis of the proposed critical residues. Availability of cDNA reagents should facilitate these efforts.

Perhaps an even greater challenge is elucidating the functional significance of N-terminal domain divergence predicted for the various intraspecies AMP deaminase polypeptides. Available nucleotide sequence information predicts that N-terminal domain divergence of related cross-species AMP deaminase variants is not random, i.e., rat A and human M are predicted to contain highly conserved N-terminal domains (Sabina et al. 1992). Since catalytic activity appears to be conferred by residues in the more highly conserved C-terminal domain of the AMP deaminase polypeptide, it is reasonable to consider non-catalytic roles for divergent N-terminal domains. Limited cellular biology studies suggest intraspecies variants of AMP deaminase occupy different intracellular addresses. In this light, it is tempting to speculate that N-terminal domain divergence is related to isoform targeting. Other possibilities, not necessarily mutually exclusive, include contributions to reported differences in kinetic/regulatory parameters and proposed heterotetramer assembly. These hypotheses are being tested by examining the cellular, molecular, and physical characteristics of chimeric AMP deaminase polypeptides created in domain swapping experiments. In conclusion, a variety of biochemical, molecular, and cellular reagents are now available which may be utilized to elucidate the functional significance of AMP deaminase isoform diversity in higher eukaryotic tissues and cells.

# References

Ashby B, Holmsen H (1981) Platelet AMP deaminase: Purification and kinetic studies. J Biol Chem 256: 10519–10523

Benders AAGM, van Kuppevelt THMSM, Oosterhof A, Veerkamp JH (1991) The biochemical and structural maturation of human skeletal muscle cells in culture: The effect of the serum substitute Ultroser G. Exper Cell Res 195: 284–294

Chang Z, Nygaard P, Chinault AC, Kellems RE (1991) Deduced amino acid sequence of *Escherichia coli* adenosine deaminase reveals evolutionarily conserved amino acid residues: Implications for catalytic function. Biochemistry USA 30: 2273–2280

Coffee CJ, Kofke WA (1975) Rat muscle 5'-adenylic acid aminohydrolase. I. Purification and subunit structure. J Biol Chem 250: 6653–6658

Cserjesi P, Olson EN (1991) Myogenin induces the myocyte-specific enhancer binding factor MEF-2 independently of other muscle-specific gene products, Mol Cell Biol 11: 4854–4862

Fishbein WN, Griffin JL, Armbrustmacher VW (1980) Stain for skeletal muscle adenylate deaminase: An effective terazolium stain for frozen biopsy specimens. Arch Pathol Lab Med 104: 462–466

Gilloteaux J, Meyer RA (1983) AMP deaminase histoenzymatic staining in rat intrafusal muscle fibers and other skeletal muscle tissues. Z Mikrosk-Anat Forsch 97: S705–S712

Kaletha K, Skladanowski A (1979) Regulatory properties of rat heart AMP deaminase. Biochim Biophys Acta 568: 80–90

Kaletha K, Spychala J, Nowak G (1987) Developmental forms of human skeletal muscle AMP-deaminase. Experientia 43: 440–443

Kaletha K, Nowak G (1988) Developmental forms of human skeletal-muscle AMP deaminase: The kinetic and regulatory properties of the enzyme. Biochem J 249: 255–261

Lowenstein JM (1972) Ammonia production in muscle and other tissues: The purine nucleotide cycle. Physical Rev 52: 382–414

Marquetant R, Desai NM, Sabina RL, Holmes EW (1987) Evidence for sequential expression of multiple AMP deaminase isoforms during skeletal muscle development. Proc Natl Acad Sci USA 84: 2345–2349

Marquetant R, Sabina RL, Holmes EW (1989) Identification of a noncatalytic domain in AMP deaminase that influences binding to myosin. Biochemistry USA 28: 8744–8749

Meyer RA, Gilloteaux J, Terjung RL (1980) Histochemical differences in AMP deaminase activity in rat skeletal muscle-fibres. Experientia 36:676–677

Meyer SL, Kvalnes-Krick KL, Schramm VL (1989) Characterization of AMD, the AMP deaminase gene in yeast. Production of *amd* strain, cloning, nucleotide sequence, and properties of the protein. Biochemistry USA 28: 8734–8743

Mineo I, Clarke PRH, Sabina RL, Holmes EW (1990) A novel pathway for alternative splicing: Identification of an RNA intermediate that generates an alternative 5' splice donor site not present in the primary transcript of *AMPD1*. Mol Cell Biol 10: 5271–5278

Mineo I, Holmes EW (1991) Exon recognition and nucleocytoplasmic partitioning determine *AMPD1* alternative transcript production. Mol Cell Biol 11: 5356–5363

Morisaki T, Sabina RL, Holmes EW (1990) Adenylate deaminase: A multigene family in humans and rats. J Biol Chem 265: 11482–11486

Morisaki T, Sabina RL, Holmes EW (1991a) A unique muscle-specific enhancer in the *AMPD1* gene. J Cell Biochem 15C(Suppl): 77

Morisaki T, Gross M, Morisaki H, Holmes E (1991b) A single mutant allele accounts for the high frequency of AMP deaminase deficiency in caucasians. Clin Res 39: 377A

Ogasawara N, Goto H, Watanabe T (1975a) Isozymes of rat AMP deaminase. Biochim Biophys Acta 403: 530–537

Ogasawara N, Goto H, Watanabe T (1975b) Isozymes of rat brain AMP deaminase: Developmental changes and characterization of five forms. FEBS Lett 58: 245–248

Ogasawara N, Goto H, Yamada Y, Yoshino M (1977) Subunit structures of AMP deaminase isozymes in rat. Biochem Biophys Res Comm 79: 671–676

Ogasawara N, Goto H, Yamada Y, Watanabe T (1978a) Distribution of AMP-deaminase isozymes in rat tissues. Eur J Biochem 87: 297–304

Ogasawara N, Goto H, Yamada Y (1978b) Effects of various ligands on interaction of AMP deaminase with myosin. Biochim Biophys Acta 524: 442–446

Ogasawara N, Goto H, Yamada Y (1979) Regulatory properties of AMP deaminase isozymes. In: Rapado A, Watts RWE, deBruyn CHMM (eds) Purine metabolism in man-III. Plenum, New York, pp 169–175

Ogasawara N, Goto H, Yamada Y, Watanabe T, Asano T (1982) AMP deaminase isozymes in human tissues. Biochim Biophys Acta 714: 298–306

Ogasawara N, Goto H, Yamada Y, Watanabe T (1984a) Distribution of AMP deaminase isozymes in various human blood cells. Int J Biochem 16: 269–273

Ogasawara N, Goto H, Yamada Y, Nishigaki I, Itoh T, Hasegawa I (1984b) Complete deficiency of AMP deaminase in human erythrocytes. Biochem Biophys Res Comm 122: 1344–1349

Pipoly GM, Nathans GR, Chang D, Deuel TF (1979) Regulation of the interaction of purified human erythrocyte AMP deaminase and the human erythrocyte membrane. J Clin Invest 63: 1066–1067

Rao N, Hara L, Askari A (1968) Alkali cation-activated AMP deaminase of erythrocytes: Properties of the membrane-bound enzyme. Biochim Biophys Acta 151: 651–654

Sabina RL, Marquetant R, Desai NM, Kaletha K, Holmes EW (1987) Cloning and sequence of rat myoadenylate deaminase cDNA: Evidence for tissue-specific and developmental regulation. J Biol Chem 262: 12397–12400

Sabina RL, Swain JL, Holmes EW (1989a) Myoadenylate deaminase deficiency. In: Scriver CR, Beaudet AL, Sly WS, Valle D (eds) The metabolic basis of inherited disease. McGraw-Hill, New York, pp 1077–1084

Sabina RL, Ogasawara N, Holmes EW (1989b) Expression of three stage-specific transcripts of AMP deaminase during myogenesis. Mol Cell Biol 9: 2244–2246

Sabina RL, Morisaki T, Clarke P, Eddy R, Shows TB, Morton CC, Holmes EW (1990) Characterization of the human and rat myoadenylate deaminase genes. J Biol Chem 265: 9423–9433

Sabina RL, Fishbein WN, Pezeshkpour G, Clarke PRH, Holmes EW (1992) Molecular analysis of the myoadenylate deaminase deficiencies. Neurology 42: 170–179

Shiraki H, Ogawa H, Matsuda Y, Nakagawa H (1979) Interaction of rat muscle AMP deaminase with myosin. I. Biochemical study of the interaction of AMP deaminase and myosin in rat muscle. Biochim Biophys Acta 566: 335–344

Stankiewicz A, Spychala J, Skladanowski A, Zydowo M (1979) Comparative studies on muscle AMP-deaminase-I. Purification, molecular weight, subunit structure and metal content of the enzyme from rat, rabbit, hen, frog and pikeperch. Comp Biochem Physiol 62B: 363–369

Stankiewicz A (1981) Human muscle AMP-deaminase. Subunit structure, amino acid composition and metal content of the homogeneous enzyme. Int J Biochem 13: 1177–1183

Thompson JL, Sabina RL, Ogasawara N, Riley DA (1992) AMP deaminase histochemical activity and immunofluorescent isozyme localization in rat skeletal muscle. J Histochem Cytochem 40: 931–946

Tovmasian EK, Hairapetian RL, Bykova EV, Severin SE, Haroutunian AV (1990) Phosphorylation of the skeletal muscle AMP-deaminase by protein kinase C. FEBS Lett 259: 321–323

Wheeler TJ, Lowenstein JM (1979) Adenylate deaminase from rat muscle: regulation by purine nucleotides and orthophosphate in the presence of 150 mM KCl. J Biol Chem 254: 8994–8999

Wilson DK, Rudolph FB, Quiocho FA (1991) Atomic structure of adenosine deaminase complexed with a transition state analog: understanding catalysis and immuno deficiency mutations. Science 252: 1278–1284

Yun S-L, Suelter CH (1978) Human erythrocyte 5'-AMP aminohydrolase: purification and characterization. J Biol Chem 253: 404–408

# 3 The Genetic Basis of Myoadenylate Deaminase Deficiency in Man

M. GROSS

Adenosine monophosphate deaminase (AMPD) catalyzes the deamination of AMP to inosine monophosphate (IMP) with the subsequent liberation of ammonia. This reaction is part of the purine nucleotide cycle (Zöllner et al. 1993). Decreased activities of the muscle isoform of AMPD (myoadenylate deaminase, MAD) in man were first described in 1962 by Pennington from muscle biopsies of five patients with Duchenne-type muscular dystrophy. Two years later, reduced activities were found in a patient with periodic hypokalemic paralysis (Engel et al. 1964). Histochemical staining for AMPD was developed in 1978. Using this technique, five patients with muscle weakness or cramping were found whose muscle biopsies were normal except for a lack of MAD activity. MAD deficiency was therefore postulated as causing a metabolic myopathy with exercise-related symptoms (Fishbein et al. 1978).

Patients with metabolic myopathy and lacking MAD activity in muscle biopsies as the only abnormal finding are considered to suffer from "primary MAD deficiency." These cases are differentiated from "secondary MAD deficiency" where AMPD as well as other muscle enzymes are not completely lacking but instead are reduced to abnormally low levels of activity due to such neuromuscular diseases as inflammatory or neurogenic myopathies (Fishbein 1985).

More than 100 cases were published within the first five years since this initial report. MAD deficiency is found in about 2%–3% of all muscle biopsies. It is the most common of all muscle enzyme defects known so far (Sabina et al. 1989).

Several isoforms of AMPD in man are encoded by at least two genes. The adult isoform of AMPD in human skeletal muscle is encoded by the AMPD1 gene (Morisaki et al. 1990; Sabina 1992).

The study described here was performed to reveal the genetic basis of MAD deficiency. The complementary (c) DNA or genomic DNA from 11 caucasian patients with primary MAD deficiency was analyzed. The frequency of the mutant allele in several populations was determined. Conclusions on the clinical significance of MAD deficiency are made possible by comparing the frequency of homozygous subjects in the population with the frequency of the deficiency in muscle biopsies.

Most of the work was performed in E. W. Holmes' laboratory at Duke University in North Carolina, USA, primarily in close cooperation with T. Morisaki and H. Morisaki. The work was supported by grants to E. W. Holmes

(DK-12413, National institutes of Health) and M. Gross (Deutsche Forschungs-gemeinschaft). Most of the findings were published by Morisaki et al. (1992). The study on the German population was conducted at the Medizinische Poliklinik München, Germany, and published by Gross et al. 1992.

## Subjects and Methods

### Subjects

Muscle biopsies from two German patients were analyzed. We are grateful to Prof. Dr. K. Gerbitz and Dr. I. Paetzke (Klinische Chemie, Städtisches Krankenhaus München-Schwabing) for providing the muscle biopsies and the results on MAD activity. Patient BA was a 25-year-old man who developed a rhabdomyolysis after viral infection. He reported no muscular symptoms other than this episode. The activity of MAD in his muscle biopsy was 18.8 U/g NCP (normal range of MAD activity in human muscle 60–300 U/g NCP, noncollagen protein). Patient MG was a 16-year-old girl who had calf pain and muscle weakness after exercise since the age of 4. The MAD activity in her muscle biopsy was 5.4 U/g NCP. Fibroblasts were cultured from 11 patients with MAD deficiency, including patient BA and patient MG.

All other patients suffered from muscle pain or weakness that were induced or aggravated by exercise. The diagnosis was based on a lack of ammonia production in the ischemic exercise test and the histochemical and biochemical examination of a muscle biopsy. Only in one patient did the case history, physical examination, and analysis of a muscle biopsy including pathological examination reveal evidence of another neuromuscular disease an addition to MAD deficiency (Becker-Kiener-type muscular dystrophy). The histological examination of all other muscle biopsies revealed no alterations or merely nonspecific changes. The mean age of the 11 patients (7 men, 4 women) was 42.8 years (range 16–68 years).

Four groups of subjects were studied additionally: 59 American caucasians, 13 Afro-Americans, 106 Japanese, and 106 Germans. None of the caucasians reported muscular symptoms. The DNA from the Afro-American and the Japanese people was provided by other laboratories without any other clinical information on the subjects.

### Methods

*Sources of DNA.* DNA was extracted from blood cells of control subjects and from cultured fibroblasts of the patients using common techniques.

*Analysis of the Muscle Biopsy of Patients BA and MG.* 60 mg of the gastrocnemius muscle were homogenized in a guanidine isothiocyanate solution and pelleted through a $CsCl_2$ gradient for RNA extraction. The reverse transcriptase reaction was carried out with 1 µg of total RNA and an AMPD1-specific 20-mer. The cDNA was amplified using the polymerase chain reaction (PCR) technique and cloned into plasmid pBS (Stratagene, CA, USA). Single-stranded plasmid DNA from pooled minipreparations was sequenced with sequenase (version 2.0) from the United States Biochemical Corporation (USB) and $^{35}$S deoxy-adenosine triphosphate (dATP) using the dideoxy chain termination method of Sanger et al. [1977] and following the recommendations of the manufacturer. Both strands of this plasmid containing the full-size cDNA were sequenced over the entire coding region of the insert.

*PCR of Exons of the AMPD1 Gene and Direct Sequencing of PCR Product.* 250 mg of *Eco* RI-digested genomic DNA were amplified using primers in the introns 1, 2, and 3, thus amplifying exon 2 and exon 3. The PCR product was sequenced by the method of Sanger et al. [1977] with $^{32}$P-γ-ATP end-labeled primers.

*Dot-Blot Analysis (Allele-Specific Oligonucleotide Hybridization).* After amplification of exon 2, PCR product was loaded onto nylon membranes (Schleicher and Schuell) and hybridized with radiolabeled allele-specific oligonucleotides that were either complementary to the wild type or mutant allele. The X-ray film was exposed overnight at −70 °C.

*MaeII digestion of PCR Product.* After amplification of exon 2, PCR product was phenol extracted and propanol precipitated. 25% of the PCR product was digested with 0.5 U *Mae*II (Boehringer Mannheim, FRG) at 50 °C for 2 h in high salt buffer. The digested DNA was size separated in a 2% agarose gel and stained with ethidium bromide.

*In Vivo Protein Expression.* The AMPD1 cDNA was expressed in the prokaryote expression vector pKK233-3 (Pharmacia, USA). *Escherichia coli* JM105 (Pharmacia, USA) was transformed with this expression vector using the calcium chloride method (Sambrook et al. 1989). Protein expression was initiated by adding isopropyl-ß-D-thiogalactopyranoside. *E. coli* lysates were assayed for AMPD activity after dialyzing against an imidazole buffer. The protein concentration was measured using the Bradford method (Sambrook et al. 1989).

*AMPD Activities Assay.* AMPD activity was assayed radiochemically ($^{14}$C-AMP from Amersham, UK) at 10mM AMP (Marquetant et al. 1987). The reaction was carried out at 37 °C in 25 mM imidazole buffer, pH 6.5, 150 mM KCl, 0.2 mg/ml bovine serum albumin (BSA) (for more details on methods see Morisaki et al. 1992).


## Results

### Analysis of Muscle Samples and DNA of MAD-Deficient Patients

Muscle biopsies of two MAD-deficient patients were analyzed. No AMPD1 peptide was detected in protein extracted from the biopsies using western blot analysis. In northern blot analysis, the AMPD1-specific mRNA was found in regular sizes and amounts. The AMPD1 cDNA was synthesized from RNA of both patients using reverse transcriptase. The cDNA was amplified using the PCR technique, cloned into plasmid pBS, and both strands of pooled plasmid preparations were sequenced completely over the entire coding region.

Two mutations were found in the cDNA: mutation $C^{34}$-T is a nonsense mutation, mutation $C^{143}$-T results in a $Pro^{48}$-Leu substitution. Nucleotide 34 is the last nucleotide in exon 2, nucleotide 143 is located in exon 3 (Fig. 1). Both mutations were confirmed in the genomic DNA of both patients. Exon 2 and exon 3 of the AMPD1 gene were amplified separately using PCR, and the amplified DNA fragments were sequenced directly for this purpose.

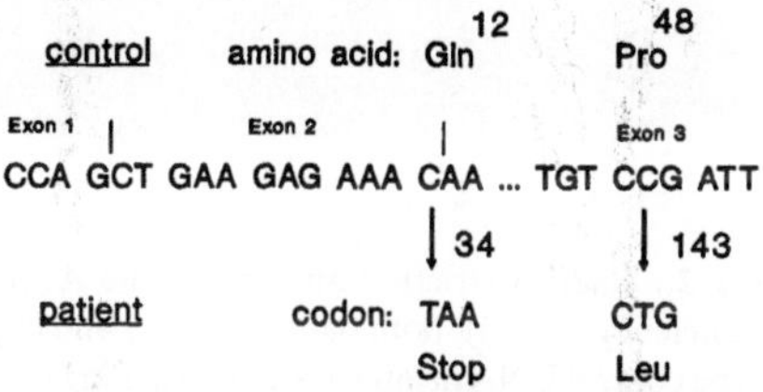

**Fig 1.** Point mutations in the AMPD1 gene of patients with MAD deficiency

The effect of the $Pro^{48}$-Leu mutation on enzyme activity was studied by in vivo expression of the mutated cDNA. A chimeric cDNA containing normal AMPD1 cDNA from nucleotides 1 to 138 (*Bgl*II restriction site) and the patient's sequence from position 139 to the 3' untranslated region was synthesized. This construct was expressed in a prokaryote expression system. No difference in specific activity was found between the enzymes containing the $Pro^{48}$-Leu mutation or the wild-type enzyme.

Nine other MAD deficient patients were screened for these two mutations. Exon 2 and exon 3 were amplified from genomic DNA using PCR and sequenced directly. All patients were homozygous for both mutations. No other mutations were found in either exon.

Genomic DNA from both parents and a brother of a patient was studied for the $C^{34}$-T mutation by *Mae*II restriction analysis: The DNA sequence at the exon 2-intron 2 boundary in the wild-type gene is 5'-AAACgtga-3' (exon 2 sequence in capitals, the *Mae*II restriction site underlined). The mutation $C^{34}$-T destroys the *Mae*II site: 5'-AAATgtga-3'. There is no other *Mae*II site between the primers used. The wild-type DNA fragment will therefore be cut by this restriction enzyme, whereas the DNA from homozygous patients will not be cut. Since the restriction site is not exactly in the middle between both primers used for PCR, two fragments of different sizes (110 and 87 base pairs ) are expected if the *Mae*II site is intact. In the case of heterozygous subjects, three bands are found: one band of uncut DNA (197 base pairs) and the two fragments of digested DNA.

The family study shows that the mutation was inherited by the heterozygous parents with spontaneous mutation being excluded (Fig. 2).

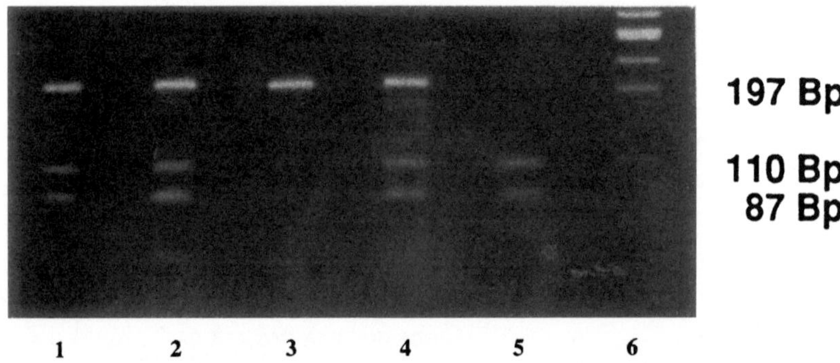


**Fig 2.** *Mae*II restriction analysis of the AMPD1 gene in a patient with primary MAD deficiency (lane 3), both parents (lane 1 and lane 2) and the patient's brother (lane 4). Wild-type control DNA in lane 5, size marker in lane 6. The three bands in lane 1,2 and 4 indicate the heterozygous status of the subjects. One band is seen in the patient, two bands in wild-type DNA

## DNA Analysis of Healthy Control Subjects

To study the frequency of the mutant allele in the population, two groups of caucasians (51 American caucasians and 106 German caucasians), 13 Afro-Americans, and 106 Japanese people were studied. Exon 2 was amplified in genomic DNA using PCR and analyzed by direct sequencing, allele-specific oligonucleotide hybridization, and/or *Mae*II digestion. The results are listed in Table 1.

**Table 1.** Frequency of the $C^{34}$-T mutation in the AMPD1 gene in several populations

| Population | $n$ | nucleotide 34 | | | Allele frequency | Expected frequency of homozygous subjects (%) |
|---|---|---|---|---|---|---|
| | | C | C/T | T | | |
| Caucasian US citizens | 51 | 39 | 10 | 2 | 0.137 | 1.9 |
| Black US citizens | 13 | 9 | 3 | 1 | 0.192 | 3.7 |
| Germans | 106 | 83 | 20 | 3 | 0.123 | 1.5 |
| Japanese | 106 | 106 | – | – | 0 | 0 |

The frequencies in the two caucasian populations match very well (13%). In contrast, the mutation was not present in 212 Japanese alleles. Based on the Hardy-Weinberg equation, 1.5%–1.9% of the Caucasian population in the United States and in Germany are expected to be homozygous for the nonsense mutation $C^{34}$-T and are therefore expected to be MAD deficient.

## Discussion

Muscle biopsies from two patients with primary MAD deficiency were analyzed. Both patients were homozygous for the nonsense mutation $C^{34}$-T and the $C^{143}$-T mutation, resulting in the amino acid substitution of proline[48] to leucine. Nine more German patients were also homozygous for both mutations. A single mutant allele therefore accounts for all cases of MAD deficiency studies thus far.

Both mutations occurred in "hot spots": The cytosine in the dinucleotide CG was substituted by thymine. This kind of mutation has been found previously in many other genes (for example hemophilia A; Youssoufian et al. 1986). It is obviously a common mutation. However, both mutations did not occur independently since all patients showed the same combination of two mutations. Both parents of one of the patients were heterozygous for the $C^{34}$-T mutation, excluding spontaneous mutations in the index patient.

There might be differences in the frequency of the mutant allele in different ethnic groups. In large numbers of muscle biopsies in the USA (Shumate et al. 1979; Heffner 1980; Fishbein et al. 1980; Kar and Pearson 1981; Kelemen et al. 1982) and Europe (Hayes et al. 1982; Joosten et al. 1982; Goebel and Bardosi 1987; Mercelis et al. 1987), 2%–3% of all muscle biopsies were MAD deficient. In contrast, not a single Oriental patient with primary MAD deficiency has thus far been reported. The report by Lally et al. [1985] on a black person is an exception. Since differences in the allele frequency seem to exist between human races, the mutation may serve as a molecular marker for the evolutionary relationship between races and ethnological groups.

Since half of the MAD-deficient biopsies are considered to be secondary deficiencies, the frequency of primary MAD deficiency might be lower than 2% of all muscle biopsies. The frequency of the mutant allele in the groups of caucasian subjects is about 13%. According to the Hardy-Weinberg equilibrium, 1.5%–1.9% of the Caucasian population are expected to be homozygous for the mutant allele. This frequency of the mutant allele in the population could easily explain the frequency of MAD deficiency in large series of muscle biopsies. This means that the $C^{34}$-T mutation is most likely the most common cause of primary MAD deficiency.

The majority of homozygous subjects obviously do not develop symptoms severe enough to cause a muscle biopsy to be taken. The percentage of MAD deficiency in muscle biopsies would otherwise be higher than the percentage of homozygous subjects in the population. Several hypotheses might explain why most homozygous subjects do not suffer from metabolic myopathy:

Exon 2 can be spliced in rat muscle without changing the reading frame (Mineo and Holmes 1991). No differences in the kinetic and regulatory properties of the enzymes encoded by this truncated or the regular cDNA were found; the biological significance of the splicing process is not yet understood. If splicing of exon 2 occurs in man, a transcript of the mutated AMPD1 gene could be produced that misses the $C^{34}$-T mutation. However, the spliced messenger (m) RNA would still contain the second mutation in exon 3. The effect of the second mutation on enzyme activity was therefore studied. After protein expression, the same specific activity was found for the mutant and wild-type enzyme. Alternative splicing of exon 2 would therefore result in a catalytically active protein even in subjects homozygous for the mutant allele. This could explain why the majority of homozygous control subjects did not suffer from a metabolic myopathy.

Another hypothesis is the expression of the AMPD2 or AMPD3 gene in asymptomatic subjects who are homozygous for the mutant allele. However, due to differences in the kinetic properties of the different isozymes, the liver or erythrocyte isoform of AMPD could hardly compensate for MAD deficiency.

Although most homozygous subjects do not develop a myopathy, their muscular performance or well-being might be affected. The performance of these persons may be reduced slightly, but not in such a manner that they complain about it. It is also possible that the majority of these persons fatigue earlier than others, although they do not undergo muscle biopsy. MAD deficiency possibly only causes symptoms in combination with other factors affecting energy metabolism. Such a factor could be the decline in mitochondrial respiratory capacity in aging (Trounce et al. 1989).

The clinical features of MAD deficiency are controversial (Zöllner et al. 1993). One of the problems is that the group of patients undergoing muscle biopsy is a selected group of symptomatic individuals. As long as diagnosis is based on muscle biopsies, asymptomatic patients or persons with mild symptoms can hardly be found at all. To further clarify these questions, we are now applying molecular genetic techniques to diagnose MAD-deficient individuals, independent of their symptoms. The results of this study may have a great impact on our knowledge about the clinical features of MAD deficiency.


# References


Engel AG, Potter CS, Rosevear JW (1964) Nucleotides and adenosine monophosphate deaminase activity of muscle in primary hypokalaemic paralysis. Nature (London) 202: 670–672

Fishbein WN (1985) Myoadenylate deaminase deficiency: inherited and acquired forms. Biochem Med 33: 158–169

Fishbein WN, Armbrustmacher VW, Griffin JL (1978) Myoadenylate deaminase deficiency: a new disease of muscle. Science 200: 545–548

Fishbein WN, Armbrustmacher VW, Griffin JL (1980) Skeletal muscle adenylate deaminase, adenylate kinase, and creatine kinase in myo-adenylate deaminase deficiency and malignant hyperthemia. Clin Res 28: 288A

Goebel HH, Bardosi A (1987) Myoadenylate deaminase deficiency. Klin Wochenschr 65: 1023–1033

Gross M, Combe C, Gresser U, Schuster H, Zöllner N (1992) Häufigkeit der für den primären Myoadenylatdeaminase-Mangel verantwortlichen Mutation C 34-T (Gin12-Stop) in einer Bevölkerungsstichprobe. Klin Wochenschr 69 (Suppl XXIII): 96

Hayes DJ, Summers BA, Morgan-Hughes JA (1982) Myoadenylate deaminase deficiency of not? Observations on two brothers with exercise-induced muscle pain. J Neurol Sci 53: 125–136

Heffner RR (1980) Myoadenylate deaminase deficiency. J Neuropathol Exp Neurol 39: 360

Joosten EMG, van Bennekom CA, Oerlemans FTJ, de Bruyn, CHMM, Oei TL, Trijbels IMF (1982) Two clinically different cases of myoadenylate deaminase deficiency: an enzyme defect in search of a disease. J Clin Chem 20: 381

Kar NC, Pearson CM (1981) Muscle adenylate deaminase deficiency. Report of six new cases. Arch Neurol 38: 279–281

Kelemen J, Rice DR, Bradley WG, Munsat TL, DiMauro S, Hogan EL (1982) Familial myoadenylate deaminase deficiency and exertional myalgia, Neurology 32: 857–863

Lally EV, Friedman JH, Kaplan SR (1985) Progressive myaglias and polyarthralgias in a patient with myoadenylate deaminase deficiency. Arthritis Rheum 28: 1298–1302

Marquetant R, Desai NM, Sabina RL, Holmes EW (1987) Evidence for sequential expression of multiple AMP deaminase isoforms during skeletal muscle development. Proc Natl Acad Sci USA 84: 2345–2349

Mercelis R, Martin JJ, de Barsy T, van de Berghe G (1987) Myoadenylate deaminase deficiency: absence of correlation with exercise intolerance in 452 muscle biopsies. J Neurol 234: 385–389

Mineo I, Holmes EW (1991) Exon recognition and nucleocytoplasmic partitioning determine AMPD1 alternative transcript production. Mol Cell Biol 11: 5356–5363

Morisaki T, Gross M, Morisaki H, Pongratz D, Zöllner N, Holmes EW (1992) Molecular basis of AMP deaminase deficiency in skeletal muscle. Proc Natl Acad Sci USA 89: 6457–6461

Morisaki T, Sabina RL, Holmes ES (1990) Adenylate deaminase. A multigene family in humans and rats. J Biol Chem 265: 11482–11486

Pennington RJ (1962) Some enzyme studies in muscular dystrophy. Proc A Clin Biochem 2: 17–18

Sabina RL (1992) The multigene family encoding AMP-deaminase deficiency in rats and humans (this volume)

Sabina RL, Swain JL, Holmes EW (1989) Myoadenylate deaminase deficiency. In: Scriver CR, Beaudet AL, Sly WS, Valle D (eds) The metabolic basis of inherited disease, 6th edn. McGraw-Hill, New York, pp 1077–1184

Sambrook J, Fritsch EF, Maniatis T (1989) Molecular cloning: a laboratory manual, 2nd edn. Cold Spring Harbor Laboratory, Cold Spring Harbor, NY

Sanger F, Nicklen S, Coulson AR (1977) DNA sequencing with chain-terminating inhibitors. Proc Natl Acad Sci USA 74: 5463–5467

Shumate J, Kaiser KK, Brooke MH, Carroll JE (1979) Myoadenylate deaminase deficiency: disease or normal variant? Neurology 29: 558

Trounce I, Byrne E, Marzuki S (1989) Decline in skeletal muscle mitochondrial respiratory chain function: possible factor in ageing. Lancet 1: 637–639

Youssoufian H, Kazazian HH, Phillips DG, Aronis S, Tsiftis G, Brown VA, Antonarakis SE (1986) Recurrent mutations in haemophilia A give evidence for CpG mutation hotspots. Nature 324: 380–382

Zöllner N, Wagner DR, Gross M (1993) Clinical aspects and biochemical basis of myoadenylate deaminase deficiency: a clinician's point of view (this volume)

# IVB  Adenylosuccinate Lyase (ASase) Deficiency

## 1  The Clinical Aspects of ASase Deficiency

J. Jaeken, P. Casaer, P. De Cock, and G. Van den Berghe

## Introduction

The discovery of adenylosuccinate lyase (EC 4.3.22; adenylosuccinase; ASase) deficiency has been the result of a new approach to investigating children with unexplained psychomotor retardation, i.e., systematic amino acid analysis of the cerebrospinal fluid (CSF) before and after strong acid hydrolysis (Jaeken 1985). In three children with severe psychomotor retardation and autistic features this technique revealed a marked increase of aspartate and glycine in CSF, only after acid hydrolysis. As a consequence of this unexpected and intriguing finding a number of additional investigations were performed on CSF including thin-layer chromatography of mono- and disaccharides which resulted in another surprise: an increase of bound ribose. Since ribose, glycine, and aspartate are building blocks of the purine skeleton the de novo purine pathway emerged as the probable site of the basic defect, more specifically at its ASase step. This hypothesis was confirmed by finding in the CSF but also in the plasma and urine of these children an accumulation of succinyladenosine (S-Ado) and succinylaminoimidazole carboxamide (SAICA) ribose, the dephosphorylated derivatives of the two substrates of ASase and by demonstrating the enzymatic defect in a number of tissues (Jaeken and Van den Berghe 1984).

Subsequently, eight other patients have been identified in Belgium and in the Netherlands (de Bree et al., 1986; Jaeken et al., 1988; Jaeken et al., 1991)

This chapter is devoted to a discussion of the clinical picture and natural course, the results of neurotechnical investigations, the concentrations of the accumulating metabolites, and the attempts at treatment.

## Clinical Picture and Natural Course

The 11 reported patients (seven females, four males) are from seven families: five families with one patient and two with three patients. They belong to four nationalities: two families are Belgian, three are Dutch, one Moroccan, and one Turkish. In the last two families there is consanguinity between the parents. The oldest living patient is 17 years (Jaeken and Van den Berghe 1984; Jaeken et al. 1988; Jaeken et al. 1991).

On clinical and biochemical grounds four variants or types can be distinguished. The majority of the patients (six from four families) have what could be called classical ASase deficiency or type I. Their clinical picture is purely neurological: severe psychomotor retardation, impressive autistic features comprising absent or poor eye contact, stereotypies, tantrums, agitation, autoaggressivity, epilepsy after the first years and axial hypotonia with normal tendon reflexes. The further evolution is characterized by absent or minimal progression of the psychomotor development and persistence of the autistic behaviour except for improved eye contact in two patients. Two patients (siblings of the Turkish family) died at ages 8 and 13 years.

The three other types are represented by only one family each. The patient with type II ASase deficiency is remarkable in that she suffers only from slight to moderate psychomotor retardation and showed a transient auditory and visual contact disturbance. She has no epilepsy. Type III, represented by three siblings of a Moroccan family, could be considered a variant of type I as the neurological features are the same. However, in addition there is a severe growth failure with muscular wasting starting between 1 and 2 years of age. One of these children died at the age of 7 years. Recently a girl was detected with a clinical symptomatology intermediate between that of type I and type II: an intermediate degree of psychomotor development, minor epilepsy and stereo-typies, aggressive behaviour and normal eye contact (type IV).

## Results of Neurotechnical Investigations

The following neurotechnical investigations yielded normal results in these patients: fundoscopy, auditory, somatosensory and visual evoked responses, nerve conduction velocities, and electromyography. It should be noted that the last two investigations, as well as the serum muscle enzymes, were also normal in the type III family with muscular wasting.

Electroencephalography was normal in infancy but mostly showed epileptic disturbances afterwards. Computed tomography and magnetic resonance imaging of the brain revealed hypotrophy (or hypoplasia?) of the cerebellum, especially of the vermis in the type I and type III patients and slight cerebral hypotrophy in the others. Positron emission tomography with 6-fluorodeoxy-glucose was performed in three patients (type I and type III) and showed a markedly reduced uptake in the cortical areas (De Volder et al. 1988).

## Concentrations of Succinylpurines in Body Fluids

In all these patients there were large increases of the dephosphorylated derivates of the two substrates of ASase, S-ado and SAICAriboside, in body fluids particularly in urine and CSF (see Table 1 for CSF levels). These com-pounds are normally undetectable. It is clear from the data in Table 1 that the values in type I and type III patients are similar, that the S-Ado concentrations in type II and type IV are three to four times higher and that the SAICA riboside levels in type IV are about twice those in the other types.

**Table 1.** Concentrations of succinylpurines and their ratios in the cerebrospinal fluid of six patients with ASase defiency (μmol/l)

| Type | S-Ado | SAICA riboside | S-Ado/SAICA riboside |
|------|-------|----------------|----------------------|
| I    | 134   | 91             | 1.5                  |
|      | 126   | 95             | 1.3                  |
|      | 133   | 120            | 1.1                  |
| II   | 475   | 128            | 3.7                  |
| III  | 162   | 95             | 1.7                  |
|      | 166   | 132            | 1.3                  |
| IV   | 379   | 214            | 1.8                  |

ASase, adenylosuccinate lyase; S-Ado, succinyladenosine; SAICA, succinylamino-imidazole carboxamide.

## Treatment

On the assumption that at least part of the symptomatology could be due to a depletion of adenine nucleotides, oral adenine (100–300 mg/day) combined with allopurinol (60–240 mg/day) was given to four type I patients for several months and to the type II patient until now, for more than 4 years. No clinical or biochemical improvement has been recorded in type I patients but there is a possible beneficial effect on growth and weight gain in the type II patient (Jaeken et al. 1988; Dr. P. Theunissen, personal communication).

## Conclusions

ASase deficiency is an autosomal recessive disorder and the first enzyme defect reported in humans along the 12-step pathway leading from phosphoribosyl-pyrophosphate to inosine monophosphate (IMP) and adenosine monophos-phate (AMP). Although the number of reported patients is rather limited, it is evident that there is usually a severe neurological involvement. However, already this small group of patients shows a remarkable biochemical and clinical heterogeneity and one of the patients has only a mild clinical disease expression.

A remarkable feature is the high incidence of autistic symptoms ("secondary autism"). An attractive hypothesis is that one of the accumulating substances (see next chapter) is responsible not only for the psychomotor retardation but also for this autism and therefore might be considered an "autistigenic" molecule. ASase deficiency could thus be a model for studying the pathogenesis of autism.

The small number of reported patients is not proof that this is a rare disorder. Indeed, the 11 patients have been found in only two medical centers (Leuven and Utrecht). Screening for this enzyme defect should systematically be performed in *all* children and adults with *any* degree of unexplained psychomotor retardation. For this purpose, a modified Bratton-Marshall test (Laikind et al. 1986) appears most practical, provided patients do not receive sulfonamides, for the measurement of which the technique was initially devised (Bratton and Marshall 1939). Finally, a systematic search for ribose in urine by thin-layer chromatography of mono- and disaccharides theoretically should be a good screening test for unraveling the many other not yet identified enzyme defects of de novo purine synthesis (Jaeken and van den Berghe 1989).

*Acknowledgements.*   This work was supported by grants 3.0026.75, 3.4653.82 and 3.4539.87 from the Fund for Medical Scientific Research (Belgium). Dr. van den Berghe is Research Director of the Belgian National Fund for Scientific Research.

# References

Bratton AC, Marshall EK (1939) A new coupling component for sulfanilamide determination. J Biol Chem 128: 537–550

de Bree PK, Wadman SK, Duran M, Fabery de Jonge H (1986) Diagnosis of inherited adenylosuccinase deficiency by thin-layer chromatography of urinary imidazoles and by automated cation exchange column chromatography of purines. Clin Chim Acta 156: 279–288

De Volder AG, Jaeken J, Van den Berghe G, Bol A, Michel C, Cogneau M and Goffinet AM (1988) Regional brain glucose utilization in adenylosuccinase-deficient patients measured by positron emission tomography. Pediatr Res 24: 238–242

Jaeken J (1985) Advances in infantile metabolic brain diseases. Ph.D. thesis, University of Leuven

Jaeken J, Van den Berghe G (1984) An infantile autistic syndrome characterized by the presence of succinylpurines in body fluids. Lancet 2: 1058–1061

Jaeken J, Van den Berghe G (1989) Screening for inborn errors of purine synthesis. Lancet 1: 500

Jaeken J, Wadman SK, Duran M, van Sprang FJ, Beemer FA, Holl RA, Theunissen PM, De Cock P, Van den Bergh F, Vincent MF, Van den Berghe G (1988) Adenylosuccinase deficiency: an inborn error of purine nucleotide synthesis. Eur J Pediatr 148: 126–131

Jaeken J, Van den Bergh F, Vincent MF, Casaer F, Van den Berghe G (1992) Adenylosuccinase deficiency: a newly recognized variant. J Inher Metab Dis 15: 416–418

Laikind PK, Seegmiller JE, Gruber HE (1986) Detection of 5'-phosphoribosyl-4-(N-succinylcarboxamide-)5-aminoimidazole in urine by use of the Bratton-Marshal reaction: identification of patients deficient in adenylosuccinate lyase activity. Anal Biochem 156: 81–90

# 2  The Biochemical Aspects of ASase Deficiency

G. Van den Berghe, F. Van den Bergh, M. F. Vincent, and J. Jaeken

## Introduction

Adenylosuccinate lyase (EC 4.3.2.2; adenylosuccinase; ASase) catalyzes two steps in the synthesis of purine nucleotides: the conversion of succinylamino-imidazole carboxamide ribotide (SAICAR) into aminoimidazole carboxamide ribotide (AICAR) along the de novo pathway, and the formation of adenosine monophosphate (AMP) from adenylosuccinate (S-AMP) in the conversion of inosine monophosphate (IMP) into adenine nucleotides (Fig. 1). Both reactions involve the cleavage of a succinyl group, yielding fumarate. ASase deficiency is the first enzyme deficiency reported in humans along the de novo pathway of purine synthesis (Jaeken and van den Berghe 1984). As reviewed in the previous chapter, the defect is transmitted as an autosomal recessive trait and results in the accumulation in body fluids of two normally undetectable compounds. SAICA riboside and succinyladenosine (S-Ado). These succinylpurines are the products of the dephosphorylation, by cytosolic 5'-nucleotidase (van den Berghe and Jaeken 1986), of the two substrates of ASase. From a clinical and biochemical investigation of eight children with ASase deficiency (Jaeken et al. 1988) two main subtypes of the defect could be distinguished: the first one, identified in seven patients and referred to as type I, is characterized by very profound psychomotor retardation, often accompanied by autistic features, and by S-Ado/SAICA riboside ratios between 1 and 2. In type II, diagnosed in one girl, mental retardation is slight, S-Ado levels are higher, and S-Ado/SAICA riboside ratios are about 4. In a third variant, neurological involvement is intermediate and higher levels of both S-Ado and SAICA riboside are found in body fluids (Jaeken et al., 1992).

In this chapter, studies of ASase in various patients' tissues and in cultured fibroblasts will be reviewed. The potential mechanisms whereby ASase deficiency may provoke mental retardation will also be discussed.

## Activities of ASase in Patients' Tissues

ASase can be measured by a dual wavelength spectrophotometric method (Schultz and Lowenstein 1976), based on the decrease in absorbance at 282 nm minus that at 320 nm, accompanying the conversion of S-AMP into AMP. The

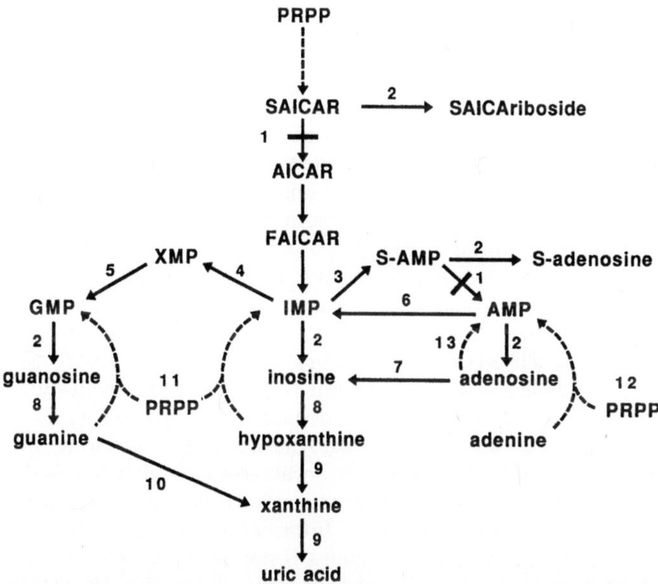


**Fig. 1.** Pathways of purine metabolism. *PRPP*, phosphoribosylpyrophosphate; *SAICAR*, succinylaminoimidazole carboxamide ribotide; *AICAR*, aminoimidazole carboxamide ribotide; *FAICAR*, formyl AICAR; *IMP*, inosine monophosphate; *S-AMP*, adenylosuccinate; *S-adenosine*, succinyladenosine; *AMP*, adenosine monophosphate; *XMP*, xanthosine monophosphate; *GMP*, guanosine monophosphate. *1.*adenylosuccinate lyase; *2*, cytosolic 5'-nucleotidase ; *3*, adenylosuccinate synthetase; *4*, IMP dehydrogenase; *5*, GMP synthetase; *6*, AMP deaminase; *7*, adenosine deaminase; *8*, purine-nucleoside phosphorylase; *9*, xanthine dehydrogenase (oxidase); *10*, guanine deaminase; *11*, hypoxanthine guanine phosphoribosyltransferase; *12*, adenine phosphoribosyltransferase; *13*, adenosine kinase. The first seven steps of the de novo synthetic pathway are represented by a *dotted line*. The adenylosuccinate lyase defect is indicated by *solid bars*.

conversion of SAICAR into AICAR can be similarly followed by the decrease in absorbance at 267 nm minus that at 320 nm (Woodward and Braymer 1966). Owing to the more than tenfold lower sensitivity of the assay with SAICAR than with S-AMP, precluding its utilization in small biopsies, inititial studies of ASase deficiency were only performed with S-AMP. More recently, we have developed a sensitive radiochemical assay to measure the activity of ASase with both its substrates on a few mg of tissue (van den Bergh et al. 1991a). It is based on the release of [14C]fumarate from either [14C]SAICAR or [14C]S-AMP. The latter are synthesized from labeled fumarate and, respectively, AICAR and AMP with partially purified ASase from yeast. Since ASase is known to be sensitive to freezing and thawing (Brand and Lowenstein 1978), assays were, as a rule, performed on fresh tissue specimens.

## Spectrophotometric Measurements

Spectrophotometric measurements of the activity of ASase with S-AMP as substrate in a series of tissues (Table 1) showed that the deficiency of the enzyme is variable and not generalized. In the liver of five type I patients, ASase activity was reduced to 15%–25% of normal; in the liver of two other patients, one with type I and the single type II patient, it was undetectable. Kidney biopsies were performed in three patients: in one of them ASase activity was undetectable, whereas in a brother and sister pair (patients 2 and 3) it was reduced to 20% of normal. The activity of muscle ASase was within the normal range in three patients; in three others (patients 2, 3, and 8) it was reduced to 10%–20% of the mean control value. Patients 2 and 3 displayed striking growth retardation (weight, height, and head circumference below the third percentiles), associated with muscular wasting, whereas in patient 8, a decrease in growth velocity had been recorded prior to therapy with adenine and allopurinol (Jaeken et al. 1988). In erythrocytes and granulocytes, ASase activity was normal. In mixed peripheral blood lymphocytes, however, it was reduced to about 40% of normal. A similar reduction of ASase activity was recorded in lymphoblasts derived from patients 2 and 3 (Barshop et al. 1989). Information concerning the activity of ASase in brain tissue of affected patients is not available. The 10- to 20-fold higher concentration of the succinylpurines in the cerebrospinal fluid than in the plasma (Jaeken and van den Berghe 1984; Jaeken et al. 1988) provides, nevertheless, a strong indication that cerebral ASase is also deficient.

**Table 1.** Activities of adenylosuccinate lyase with S-AMP

| | Liver | Kidney | Muscle | RBC | WBC | PBL |
|---|---|---|---|---|---|---|
| Controls | $0.72 \pm 0.18$ | $1.08 \pm 0.10$ | $2.61 \pm 0.34$ | $1.20 \pm 0.16$ | $1.63 \pm 0.11$ | $1.27 \pm 0.14$ |
| Type I patients | | | | | | |
| 1 | 0.10 | 0 | 2.77 | 0.96 | 1.37 | 0.50 |
| 2 | 0.14 | 0.18 | 0.28 | 0.82 | – | 0.48 |
| 3 | 0.11 | 0.21 | 0.66 | 0.73 | 1.22 | 0.42 |
| 5 | 0.13 | – | 2.18 | – | – | – |
| 6 | 0 | – | – | – | – | – |
| 7 | 0.19 | – | 1.28 | – | – | – |
| Type II patient | | | | | | |
| 8 | 0 | – | 0.39 | – | – | – |

Activities were measured spectrophotometrically with 0.2 m$M$ adenylosuccinate (S-AMP) and are expressed as nmol/min per mg of protein, except in red blood cells (RBC) for which nmol/min per mg of Hb was used. Control values represent means $\pm$ SEM for 5–8 samples. Patient numbers correspond to those given in Jaeken et al. (1988). WBC, granulocytes; PBL, mixed peripheral blood lymphocytes; –, not determined.

Taken together, the enzyme studies indicate the existence of isoforms of ASase and of genetic heterogeneity of the enzyme defect. That isozymes of mammalian ASase exist is also apparent from the observation that starvation induced a profound decrease of the activity of ASase in rat liver and spleen, but had no effect on its activity in brain, muscle, and kidney (Brand and Lowenstein 1978). These isozymes have, however, not yet been characterized.

## Radiochemical Measurements

Activities of ASase were measured radiochemically with both substrates in a number of liver biopsies which had been stored at $-80\,°C$ for approximately 5 years (Table 2). In control biopsies, about 20% of the enzyme activity, measured with S-AMP in the fresh state, had been lost by freezing and storage (compare with Table 1). In these samples, the ratio of the activity measured with SAICAR compared to that with S-AMP was about 0.4. This compares with a ratio of 0.7 reported for purified ASase from human erythrocytes (Barnes and Bishop 1975). In samples of type I patients 1–3, ASase activities with S-AMP were closely similar to those measured previously. In addition, they were comparable to that obtained in a fresh unfrozen fragment taken by needle biopsy from a newly diagnosed patient (number 9, a younger sister of patients 2 and 3). In all the patients' samples, activities with SAICAR were reduced to the same extent as those with S-AMP, namely to 15%–20% of normal. The ratio of the activity measured with SAICAR compared to that with S-AMP was between 0.4 and 0.7 in these samples, indicating that both activities were lost in parallel.

In deep-frozen human control muscle and kidney, an approximately 50% loss of activity was recorded when the values measured with S-AMP in 5-year-old sam-

**Table 2.** Activities of liver adenylosuccinate lyase with S-AMP and SAICAR

|          | S-AMP           | SAICAR          | Ratio           |
|----------|-----------------|-----------------|-----------------|
| Controls | $0.57 \pm 0.04$ | $0.24 \pm 0.03$ | $0.42 \pm 0.02$ |
| Patients |                 |                 |                 |
| 1        | 0.12            | 0.05            | 0.4             |
| 2        | 0.10            | 0.04            | 0.4             |
| 3        | 0.12            | 0.06            | 0.5             |
| 9        | 0.09            | 0.06            | 0.7             |

Activities were measured radiochemically with 0.2 m$M$ substrate and are expressed as nmol/min per mg of protein. Samples from controls (means $\pm$ SEM for $n = 3$) and from patients 1–3 (numbers correspond to those given in Jaeken et al. 1988) had been kept at $-80\,°C$ for 5 years. Activities in patient 9 (a younger sister of patients 2 and 3) were measured on fresh, unfrozen tissue. S-AMP, adenylosuccinate; SAICAR, succinylaminoimidazole carboxamide ribotide

ples were compared with those obtained on fresh tissue (results not shown). In these samples, the ratio of activity with SAICAR compared to that with S-AMP was also between 0.4 and 0.7. The same ratio was recorded in deep frozen samples from type I ASase deficient patients, indicating that both activities of ASase were also lost in parallel in these tissues. This parallel loss of activity accords with the observation that the dephosphorylated derivatives of the two substrates of the enzyme accumulate in approximately equimolar amounts in type I patients. Unfortunately, biopsies of the type II patient were no longer available for determination of the activity of ASase with its two substrates.

## Studies of Normal and Mutant ASase in Fibroblast Extracts

In cultured fibroblasts from the three initially identified, type I ASase deficient patients, a partial deficiency of the enzyme, measured with S-AMP, was found (van den Berghe and Jaeken 1986), associated with a decrease of its stability (Laikind et al., 1986). Recently, using the radiochemical assay (van den Bergh et al. 1991a), we have performed detailed studies of the activity of ASase with its two substrates in normal cultured fibroblasts and of the residual activity of the enzyme in cells taken from several affected children. These have revealed striking differences between type I patients and the single type II patient (van den Bergh et al. 1991b).

### Kinetic Parameters

In the fibroblasts of type I patients, $V$max of ASase with S-AMP and with SAICAR were both decreased to about 30% of the respective control values (Table 3). In the type II patient, $V$max with SAICAR was also approximately 30% of control but, in marked contrast with the other patients, that with S-AMP was reduced to 3% of normal. The much more pronounced loss of activity with S-AMP as compared to SAICAR, if also present in other tissues, provides an explanation for the higher S-Ado/SAICA riboside ratio (4–5 vs 1–2) in the body fluids of the type II patient.

The apparent $K$m of S-AMP for fibroblast ASase from type I patients was not modified as compared to control cells (Table 3). Similar results were obtained with ASase from cultured lymphoblasts derived from these patients (Barshop et al. 1989). Owing to the very low residual activity with S-AMP in the cells from the type II patient, $K$m of S-AMP could not be measured. In both mutant cell types, the $K$m of SAICAR was increased two-to-fourfold as compared to control cells, although low activities hampered precise measurements. These data indicate that, although $V$max was decreased similarly with both S-AMP and SAICAR in type I cells, the activity with SAICAR was more affected by the mutation than that with S-AMP. Moreover, the results obtained with the fibroblasts of the type II patient confirm the clinical observation that this child carries a different mutation than the majority of the ASase deficient patients.

**Table 3.** Kinetic parameters of adenylosuccinate lyase activities in control and in patient fibroblasts

| | $V_{max}$ (nmol/min/mg protein) | | $K_m$ ($\mu M$) | |
|---|---|---|---|---|
| | *S-AMP* | *SAICAR* | *S-AMP* | *SAICAR* |
| Controls | $1.44 \pm 0.1$ | $0.92 \pm 0.20$ | $10 \pm 2$ | $7 \pm 2$ |
| Type I Patients | | | | |
| 1 | 0.44 | 0.22 | 10 | ~30 |
| 2 | 0.46 | 0.31 | 11 | ~30 |
| 3 | 0.43 | 0.22 | 9 | ~20 |
| 7 | 0.47 | 0.26 | 11 | ~15 |
| Type II patient | | | | |
| 8 | 0.04 | 0.31 | – | ~25 |

Activities were measured radiochemically and are expressed as nmol/min per mg of protein. $V$max was determined at 0.2 m$M$ substrate Control values represent means $\pm$ SEM for three cell lines. Patient numbers correspond to those given in Jaeken et al. (1988); S-AMP, adenylosuccinate; SAICAR, succinylaminoimidazole carboxamide ribotide; –, non measurable.

## Effects of Salts and Nucleotides

In control cells, the activity of ASase, measured both with S-AMP and SAICAR, was not influenced by the addition of 100 m$M$ KCl. Similar results were obtained in cells of type I patients. In contrast, KCl markedly inhibited ASase activity with SAICAR in type II fibroblasts (activity decreased by approximately 60% with 100 m$M$ KCl). Further studies showed that this inhibition was due to the anion, competitive ($K$m increased to 270 $\mu M$ in the presence of 90 m$M$ KCl), and also exerted by other anions which inhibited in the following order: $KH_2PO_4 > K_2SO_4 > KI > KBr > KCl > KF > KCOOH$.

In cells from controls and from type I patients the activity with SAICAR was not inhibited by purine and pyrimidine nucleoside triphosphates. In contrast, the activity of ASase in cells of the type II patient was markedly (60%–90%) inhibited by 2.5 m$M$ adenosine triphosphate (ATP), guanosine triphosphate (GTP), inosine triphosphate (ITP), aminoimidazole carboxamide ribotide triphosphate (ZTP), uridine triphosphate (UTP), and cytidine triphosphate (CTP). All these results corroborate the conclusion that the mutation in the type II patient markedly differs from that in type I patients.

 

# Studies in Intact ASase Deficient Fibroblasts

Fibroblasts from both the type I patients and the type II patient displayed growth curves that were similar to those of control cells in medium containing undialyzed fetal calf serum. This may be explained by the observation that, with the exception of the activity with S-AMP in type II cells, cultured fibroblasts have only a partial ASase deficiency. Fibroblasts also synthesize purine nucleotides from purine bases via the salvage pathway. In medium prepared with dialyzed serum, in which purine bases such as hypoxanthine have been removed, growth of control fibroblasts was distinctly reduced, reflecting decreased synthesis of purine nucleotides via the salvage pathway. However, cells from ASase deficient patients tended to grow more rapidly than control cells, particularly after 12 days of culture, suggesting an adaptation to the enzyme defect. To study this adaptation, experiments were performed in which the incorporation of labeled precursors into fibroblast purine nucleotides was measured. These studies also allowed us to evaluate the functional consequences of decreased ASase activity on the conversion of S-AMP into AMP and of SAICAR into AICAR.

## Conversion of S-AMP into AMP

The conversion of S-AMP into AMP can be evaluated by following the incorporation of labeled hypoxanthine into the adenine nucleotides. In fibroblasts, which lack xanthine dehydrogenase, hypoxanthine is exclusively utilized by hypoxanthine guanine phosphoribosyltransferase (HGPRT) and thereby converted into IMP (Fig. 1). IMP can be converted either into guanine nucleotides via IMP dehydrogenase and GMP synthetase or into adenine nucleotides via adenylosuccinate synthetase and ASase. Radioactivity in the adenine nucleotides thus reflects flux through ASase. Rates of incorporation of 20 $\mu M$ [$^{14}$C]hypoxanthine into total purine nucleotides measured over 30 min in fibroblasts of ASase deficient patients, resuspended in Krebs-Ringer bicarbonate buffer, were comparable to those measured in control cells, indicating similar activities of HGPRT. In both control fibroblasts and in cells from type I patients, about 5% of incorporated radioactivity was recovered in IMP, 7% in guanine nucleotides, and 88% in adenine nucleotides after 20 min. Neither S-AMP nor S-Ado were detected. In contrast, in the cells of the type II patient, 9% of incorporated radioactivity was recovered in IMP, 7% in guanine nucleotides, only 25% in adenine nucleotides, but 56% in S-AMP, and 3% in S-Ado. This indicates that only the pronounced deficiency of ASase activity with S-AMP hampers metabolic flux through the enzyme step. Nevertheless, even the small residual activity recorded (approximately 3% of normal, Table 3) still allows conversion of S-AMP into AMP and the other adenine nucleotides to proceed. The limited amount of S-Ado formed is in accordance with previous studies which have shown a low rate of dephosphorylation of S-AMP by purified cytosolic 5'-nucleotidase from rat liver (van den Berghe and Jaeken 1986).

**Conversion of SAICAR into AICAR**

This conversion can be evaluated by following the incorporation of formate into the total purine nucleotide pool. Formate is utilized in the conversion of glycineamide ribotide (GAR) into formylglycineamide ribotide (FGAR) (not shown on Fig. 1) and of AICAR into formyl AICAR (FAICAR) (Fig. 1), respectively the third and ninth steps of de novo synthesis of IMP. Although formate incorporation also proceeds distally from ASase, this incorporation is dependent on the generation of AICAR by ASase. Insofar as conversion of SAICAR into AICAR remains possible, incorporation of formate into the adenine nucleotides will also occur if conversion of S-AMP into AMP is possible. The rate of incorporation of 0.2 m$M$ [$^{14}$C]formate into total purine nucleotides, measured over 7 h in cells grown as monolayers in medium containing dialyzed fetal calf serum, was about 40% higher in cells of type I patients and 60% higher in cells of the type II patient than in control fibroblasts. This is in accordance with previous work (Laikind et al. 1986) and indicates an increased de novo synthesis of purine nucleotides in the ASase deficient cells. The mechanism of this increase remains to be determined. After 7 h, both in control fibroblasts and in cells from type I patients, approximately 90% of incorporated radioactivity was in the adenine nucleotides and the remainder in the guanine nucleotides. Labeled SAICAR and S-AMP could not be detected. These results indicate that the partial defect of ASase activity with SAICAR in type I cells does not hinder flux through the de novo pathway. In contrast, in cells from the type II patient, 6% of incorporated radioactivity was recovered in SAICAR and 5% in S-AMP. This indicates that the partial defect of ASase activity with SAICAR in type II cells, although of similar magnitude as in type I (about 30% residual activity; see Table 3), impedes conversion of SAICAR into AICAR. This could be explained by the inhibitory effect of salts and nucleotides on ASase activity in the type II fibroblasts and by the higher flux through the de novo pathway recorded in these cells, which may render conversion of SAICAR into AICAR limiting. The accumulation of radioactive S-AMP upon incubation with labeled formate confirms the results obtained with labeled hypoxanthine.

# Potential Pathophysiological Mechanisms in ASase Deficiency

Several hypotheses can be put forward to explain the symptoms of ASase deficiency. As in all inborn errors of metabolism, deficiencies of metabolites which are normally formed distally from the enzyme defect, as well as accumulations of intermediates proximally thereof, could have deleterious effects. Mutations leading to a total loss of ASase in all tissues are probably incompatible with life. They would suppress de novo synthesis of purines and render cells completely dependent on the purine salvage enzymes, HGPRT, adenine phosphoribosyltransferase (APRT), and adenosine kinase (Fig. 1). Moreover, synthesis of adenine nucleotides would depend exclusively on APRT and adenosine kinase, since ASase also intervenes in the salvage of hypoxanthine into AMP. That the purine salvage enzymes could sustain normal purine nucleotide synthesis on their own is unlikely.

## Deficiencies of Purine Nucleotides

From the dual function of ASase, one would have expected its deficiency to lead to decreased concentrations of purine and particularly of adenine nucleotides. However, measurements of the concentrations of adenine and guanine nucleotides in freeze-clamped liver and muscle of ASase deficient patients (van den Berghe and Jaeken 1986) and $^{31}$P magnetic resonance spectroscopy of nucleotide phosphates in their brains (Dorland et al. 1986) have yielded normal results. This suggests compensation of the ASase defect by supply of purines from nonaffected cell types (erythrocytes, granulocytes) via the purine salvage enzymes HGPRT, APRT, and adenosine kinase. Another possibility is that, as in cultured fibroblasts, residual ASase activity remains sufficient to allow the required flux through the pathway of purine synthesis. Nevertheless, a deficiency of purine nucleotides could occur in some cell types with profound deficiency of ASase and low activities of the salvage enzymes.

Recent studies have shown that AICAR (now often called ZMP), the product of the ASase reaction with SAICAR, can be converted into a triphosphoderivative, ZTP. Accumulation of ZTP has been documented in red blood cells of HGPRT deficient patients (Sidi and Mitchell 1985), and upon treatment of various cell types with AICA riboside (Sabina et al. 1985; Vincent et al. 1991). Whether ZTP is found in normal cells under physiological conditions and whether it is an essential cellular constituent remain open questions. However, if the answers were positive, ASase deficiency could induce a deleterious ZTP depletion.

## Impairment of the Purine Nucleotide Cycle

The purine nucleotide cycle is composed of the three enzymes, adenylosuccinate synthetase, ASase, and AMP deaminase. It has been reported to be particularly active in muscle and to account for ammonia production during muscle contraction (reviewed in Lowenstein 1972). Impairment of the function of the purine nucleotide cycle could thus play a role in the pathogenesis of the growth retardation and muscle wasting of the patients in whom the defect is expressed in muscle. Although the purine nucleotide cycle may also operate in brain under ischemic situations, its relationship with brain function in normal conditions and the consequences of its interruption on psychomotor development remain to be established.

## Accumulation of SAICAR and S-AMP

Accumulation of both substrates of ASase was undetectable in liver and muscle, but a slight buildup of S-AMP could be measured in the kidney of ASase deficient patients (van den Berghe and Jaeken 1986). This might be explained by more efficient dephosphorylation of SAICAR and S-AMP in liver and muscle than in the kidney. Nevertheless, the possibility exists that small intracellular accumulations of the substrates of Asase, remaining below or at the limit of the level of detection of the methods used, may have deleterious effects in some cell types.

## Accumulation of SAICA Riboside and S-Ado

Owing to the resemblance of both succinylpurines to adenosine and to their accumulation to 100–500 $\mu M$ concentrations in the cerebrospinal fluid of the patients, the possibility was explored that they might interfere with cerebral adenosine receptors and thereby with the numerous physiological functions of adenosine. In the central nervous system, these include vasodilation, sedation, and inhibition of neurotransmitter release and nerve cell firing (reviewed by Dunwiddie 1985). Studies with crude membrane fractions of rat cerebral cortex, however, failed to show interference of SAICA riboside and S-Ado with binding and uptake of adenosine (Vincent and van den Berghe 1989).

Measurements of the cerebral uptake of [6-$^{18}$ $F$2-deoxyglucose by positron emission tomography have shown that it is markedly reduced in the cortical areas of ASase deficient patients (de Volder et al. 1988), suggesting interference of the succinylpurines with glucose metabolism. However, when SAICA riboside and S-Ado were tested on glucose metabolism in the isolated hepatocyte model, no effects were seen (M.F. Vincent and G. van den Berghe, unpublished data).

# Concluding Remarks

Although substantial progress has been made regarding our knowledge of ASase deficiency, including its characterization at the gene level, described by Stone et al. in the next chapter, much work remains to be done, particularly with respect to the mechanisms whereby the defect exerts its deleterious effects on brain function. From the observation of a strikingly less severe psychomotor retardation in the single patient with high S-Ado/SAICA riboside ratios, it is tempting to conclude that SAICA riboside is the offending compound and that S-Ado could protect against its toxic effects. Further studies should thus be directed at searching for the effects of SAICA riboside on brain metabolism and physiology.

Additional work should also be performed on the structure of normal and genetically modified ASase. Mammalian ASase is composed of four subunits with a molecular weight of approximately 52K (Casey and Lowenstein 1987). The observations in the ASase deficient patients suggest that the native enzyme in some tissues may be a heteropolymer formed from subunits encoded by different genes. The mutation in type I patients might have affected one species of subunit, whereas that in the type II patient might have altered another. Alternatively, in the latter patient, the mutation might have affected differently the binding of S-AMP and of SAICAR to the enzyme.

*Acknowledgements.*   This work was supported by grant 3.4539.87 of the Fund for Medical Scientific Research (Belgium) and by the Belgian State – Prime Minister's Office for Science Policy Programming. G. van den Berghe is Director of Research of the Belgian National Fund for Scientific Research.

# References

Barnes LR, Bishop SH (1975) Adenylosuccinate lyase from human erythrocytes. Int J Biochem 6: 497–503

Barshop BA, Alberts AS, Gruber HE (1989) Kinetic studies of mutant human adenylosuccinase. Biochim Biophys Acta 999: 19–23

Brand LM, Lowenstein JM (1978) Effect of diet on adenylosuccinase activity in various organs of rat and chicken. J Biol Chem 253: 6872–6878

Casey PJ, Lowenstein JM (1987) Purification of adenylosuccinate lyase from rat skeletal muscle by a novel affinity column. Stabilization of the enzyme, and effects of anions and fluoro analogues of the substrate. Biochem J 246: 263–269

De Volder AG, Jaeken J, van den Berghe G, Bol A, Michel C, Cogneau M, and Goffinet AM (1988) Regional brain glucose utilization in adenylosuccinase-deficient patients measured by positron emission tomography. Pediatr Res 24: 238–242

Dorland L, van Sprang FJ, van Echteld CJA, Duran M, Wadman SK, den Hollander JA, Luyten PR (1986) In vivo magnetic resonance spectroscopy and imaging of patients with adenylosuccinase deficiency. Annual Meeting of SSIEM, Abstract book P 150. Amersfoort, The Netherlands

Dunwiddie TV (1985) The physiological role of adenosine in the central nervous system. Int Rev Neurobiol 27: 63–139

Jaeken J, van den Berghe G. (1984) An infantile autistic syndrome characterised by the presence of succinylpurines in body fluids. Lancet 2: 1058–1061

Jaeken J, Wadman SK, Duran M, van Sprang FJ, Beemer FA, Holl RA, Theunissen PM, de Cock P, van den Bergh F, Vincent MF, van den Berghe G (1988) Adenylosuccinase deficiency: an inborn error of purine nucleotide synthesis. Eur J Pediatr 148: 126–131

Jaeken J, van den Bergh F, Vincent MF, Casaer P, van den Berghe G (1992) Adenylosuccinase deficiency: a newly recognized variant. J Inher Metab Dis 15: 416–418

Laikind PK, Gruber HE, Jansen I, Miller L, Hoffer M, Seegmiller JE, Willis RC, Jaeken J, van den Berghe G (1986) Purine biosynthesis in chinese hamster cell mutants and human fibroblasts partially deficient in adenylosuccinate lyase. Adv Exp Med Biol 195B: 363–369

Lowenstein JM (1972) Ammonia production in muscle and other tissues: the purine nucleotide cycle. Physiol Rev 52: 384–414

Sabina RL, Patterson D, Holmes EW (1985) 5-Amino-4-imidazolecarboxamide riboside (Z-riboside) metabolism in eukaryotic cells. J Biol Chem 260: 6107–6114

Schultz V, Lowenstein JM (1976) Purine nucleotide cycle. Evidence for the occurrence of the cycle in brain. J Biol Chem 251: 485–492

Sidi Y, Mitchell BS (1985) Z-nucleotide accumulation in erythrocytes from Lesch-Nyhan patients. J Clin Invest 76: 2416–2419

Van den Bergh F, Vincent MF, Jaeken J, van den Berghe G (1991a) Radiochemical assay of adenylosuccinase: demonstration of parallel loss of activity toward both adenylosuccinate and succinylaminoimidazole carboxamide ribotide in liver of patients with the enzyme defect. Analyt Biochem 193: 287–291

Van den Bergh F, Vincent MF, Jaeken J, van den Berghe G (1991b) Adenylosuccinase activity and succinylpurine production in fibroblasts of adenylosuccinase-deficient children. Adv Exp Med Biol 309B: 277–280

Van den Berghe G, Jaeken J (1986) Adenylosuccinase deficiency. Adv Exp Med Biol 195A: 27–33

Vincent MF, van den Berghe G (1989) Influence of succinylpurines on the binding of adenosine to a particular fraction of rat cerebral cortex. Adv Exp Med Biol 253B: 441–445

Vincent MF, Marangos PJ, Gruber HE, van den Berghe G (1991) Inhibition by AICAriboside of gluconeogenesis in isolated rat hepatocytes. Diabetes 40: 1259–1266

Woodward DD, Braymer HD (1966) Purification and properties of Neurospora adenylosuccinase. J Biol Chem 241: 580–587

# 3    The Genetic Basis of ASase Deficiency

R. L. Stone, J. Aimi, B. A. Barshop, J. Jaeken, G. Van den Berghe, H. Zalkin, and J. E. Dixon

## Isolation of the First Vertebrate Clones Encoding the Enzymes of De Novo Purine Synthesis

### Functional Complementation of Microbial Auxotrophs

During the past 2 years, our goal has been to clone all of the enzymes of vertebrate de novo purine synthesis by functional complementation of microbial auxotrophic mutants. We employed an avian liver cDNA expression library excised as phagemids from the bacteriophage λ vector λZAP-II. We reasoned that the mRNA levels in avian liver would be elevated because, lacking a urea cycle, avian systems use purine biosynthesis for ammonia excretion. We and others have isolated vertebrate cDNAs encoding 11 of the 14 activities in the pathway by this methodology [1–5]. With this method, we were unable to isolate clones encoding the first and fourth enzymatic steps leading from PRPP to IMP. We were also unable to isolate a clone encoding adenylosuccinate synthetase.

### Polymerase Chain Reaction Cloning

Using degenerate oligonucleotide primers based upon conserved amino acid sequences within the microbial enzymes, our laboratory has been able to isolate a cDNA clone encoding the first activity leading from PRPP to IMP and a cDNA clone encoding adenylosuccinate synthetase. In each case, the clones encoded peptides with $NH_2$-terminal extensions that required removal to yield an active enzyme in the corresponding *E. coli* auxotroph.

### Evolution of Multifunctional Enzymes in Vertebrates

One result of our cloning of the vertebrate enzymes of de novo purine synthesis has been the discovery of multifunctional vertebrate enzymes. A trifunctional vertebrate enzyme houses the second, third, and fifth activities of the pathway, while bifunctional enzymes catalyze the sixth and seventh and the ninth and tenth activities. In prokaryotes, each of these activities is catalyzed by a distinct protein.

# The Vertebrate Adenylosuccinate AMP-Lyase cDNAs

## The Avian Clone

One of the avian cDNA clones obtained by functional complementation encodes adenylosuccinate AMP-lyase (ASL). This enzyme catalyzes the eight and
twelfth steps of the pathway leading from PRPP to AMP. Since the two substrates of the enzymes are very closely related in chemical structure, this enzyme
is believed to contain only one active site and is, therefore, not considered to be
a bifunctional enzyme.

## The Human Clone

Using low stringency hybridization conditions with the avian cDNA serving as
a probe, we were able to isolate a cDNA clone of human liver ASL. This clone
shared 85% amino acid identity with the avian clone and lesser degrees of
identity with the *B. subtilis* and *E. coli* enzymes (Fig. 1). Among the amino acid
residues which are absolutely conserved is a fumarate lyase signature sequence,
common to all enzymes catalyzing β-elimination reactions which liberate fumarate as a product. Also conserved is a histidyl residue believed to function in
catalysis (see [4]).

## Deficiency in a Moroccan Family

We have focused our research on human ASL because a deficiency in this
enzyme's activity has been documented in ten cases of severe psychomotor
retardation [6–8]. The studies discussed here have centered on the three
affected siblings of a Moroccan family. The grandparents of the children are first
cousins. The parents are reported to be normal and healthy and there are four
unaffected children in the family.

# The ASL cDNA of Moroccan Siblings with Psychomotor Retardation

## PCR Strategy

We obtained Epstein-Barr virus-transformed lymphoblast lines from two affected siblings in the family [9] and isolated total cellular RNA. Using four sets
of primers based on the human liver cDNA sequence, overlapping PCR clones
of ASL cDNA were generated from randomly primed first strand cDNA syn-
theses from each RNA source. The PCR products were subcloned and subjected
to nucleotide sequence analysis.

                                                          R. L. STONE et al.

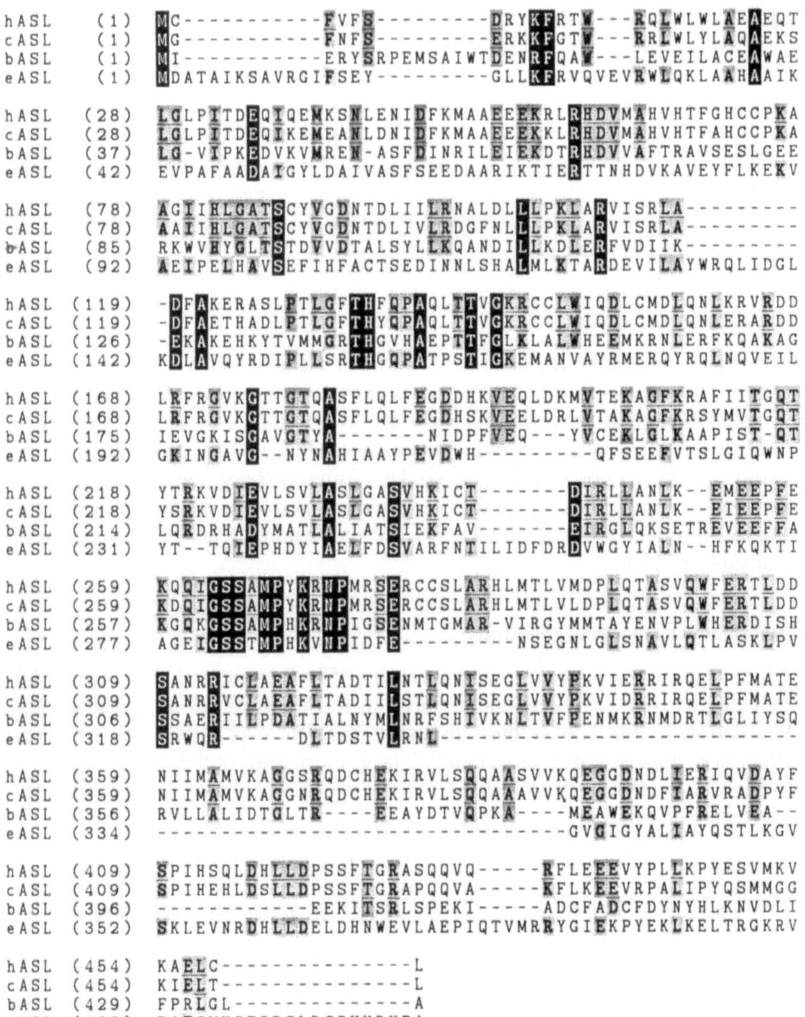



**Fig. 1.** Amino acid sequence comparison of human ASL (*hASL*), chicken ASL (*cASL*), *B. subtilis* ASL (*bASL*) and *E. coli* ASL (*eASL*). The amino acid sequences of the four enzymes were aligned to yield maximum positional identity. Absolute identity at a given position is indicated by *black boxes*. Identity at a given position in three out of four of the enzymes is indicated by a *gray box*. In this figure all acidic amino acids are considered to be identical to one another, as are the basic amino acids. Amino acid numbers are indicated for each enzyme.

## The Point Mutation

In the ASL cDNAs of each affected sibling, a single nucleotide change was detected. This point mutation resulted in the change of serine residue 413 to proline. The point mutation also created a new *Hph*I restriction endonuclease recognition site in the cDNA (Fig. 2). In all, 12 distinct subclones representing four distinct PCR reactions were sequenced from each patient. The point mutation was the only change found in both the coding and noncoding regions of the cDNA.

**Fig. 2.** The mental retardation-associated point mutation. The figure diagrams a PCR product spanning the C-terminal one fourth of the adenylosuccinate lyase cDNA. As shown in the figure, the human liver cDNA contains two *Hph*I recognition sequences and a TCC codon as codon number 413. The PCR products from the affected siblings have a T to C transition, changing codon 413 to CCC and the encoded amino acid from serine to proline. The point mutation also creates a new *Hph*I recognition site

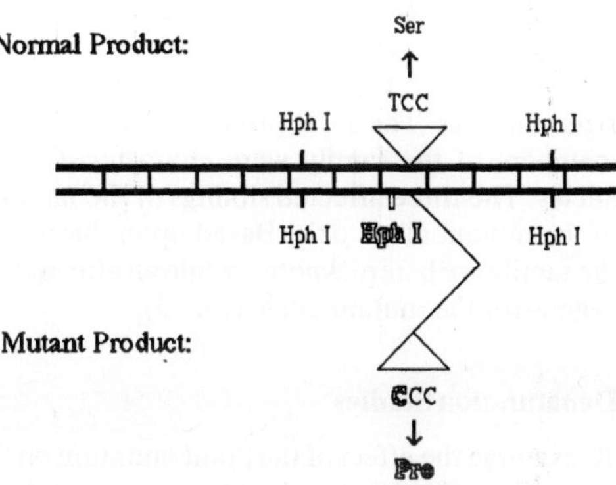


## Controls

The point mutation changes an amino acid residue which is not absolutely conserved between species. This lead to the concern that the mutation might represent a variant of ASL which is found relatively frequently in the normal human population. To examine this possibility we used genomic DNA from 55 control subjects as a substrate for the generation of PCR products from the region of the ASL gene containing the mutation. The products generated from genomic DNA were the same size as those expected from the cDNA. *Hph*I digestion of the PCR products generated the fragment pattern expected from the normal ASL allele. This result indicated that the point mutation does not occur at a high frequency in the normal human population.

## Pedigree of the Family

Convinced that the point mutation was not a frequently occurring variant of ASL, we turned our attention to the segregation of the point mutation with the developmental disorder in the family studied. PCR products form the DNA of family members were generated in the region of the mutation and subjected to

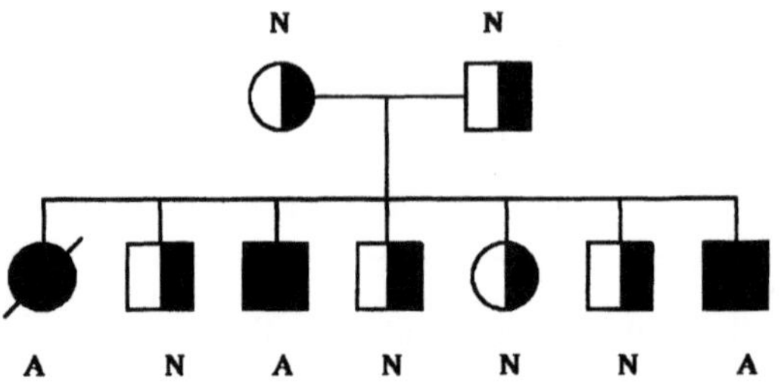



**Fig. 3.** Pedigree of the family studied. In the figure, *circles* indicate males and *squares* indicate females. The children in the family are presented in birth order from left to right. Normal (*N*) and affected (*A*) individuals are indicated. The presence of the point mutant allele is indicated by *dark shading*. All affected members of the family are heterozygotes. The first affected child in the family is deceased as indicated by the *slash mark*

*Hph*I analysis. The PCR products derived from the parents and all unaffected members of the family were characteristic of both the normal and mutant alleles. The three affected siblings of the family yielded products characteristic of the mutant allele only. Based upon this analysis, all unaffected members of the family are heterozygotes, while all affected members of the family are homozygous for the mutant allele (Fig. 3).

## Denaturation Studies

To examine the effect of the point mutation on the structure of the ASL enzyme, we expressed the normal and mutant proteins in *E. coli* using the pT7–7 expression system of Tabor and Richardson [10]. We found no differences in the two enzymes in their affinity for S-AMP. However, we found that the mutant enzyme's activity was much more sensitive to denaturation by urea and guanidine-HCl than was that of the normal enzyme isolated under the same conditions (Fig. 4). The change in free energy of the active site structure caused by the point mutation ($\Delta\Delta G_{[H_2O]}$ = 0.2 kcal/mol) implies that a few critical hydrophobic interactions are lost in the mutant.

## Steady State Enzyme Levels In Situ

To assess the relationship of the in vitro lability of the mutant ASL to the steady state level of ASL in vivo, we performed western analysis. In lymphoblast cell lines derived from two of the affected siblings, the steady state level of ASL was markedly decreased in comparison to the control cells, while the lactate dehydrogenase levels in all three cell lines were identical. This may suggest that the point mutation leads to a greater rate of degradation within the cell. The half-life of normal and mutant ASL is currently being assessed.

**Fig. 4** Fractional activity of recombinant human ASL in response to denaturation. In all panels the *open circles* and *solid lines* indicate the curve obtained with the normal enzyme, while the *open squares* and *dashed lines* indicate the mutant enzyme described in the text. **a** Partially purified samples of normal and mutant human ASL were treated with varying concentrations of guanidine-HCl (abscissa) for 1 hour at room temperature in assay buffer (10 m$M$ Tris-HCl, pH 7.4; 10 m$M$ KCl, 2 m$M$ EDTA, 1 m$M$ DTT). ASL activity was then assessed and compared to the activity of undenatured enzyme maintained at room temperature in the same buffer. Longer periods of preincubation did not alter the results of the assays, indicating that equilibrium was reached in 1 h. Further, dilution of a higher concentration of denaturant to a lower concentration resulted in the fractional activity expected for the final denaturant concentration, indicating that the denaturation monitored in this experiment is reversible. **b** Partially purified samples of normal and mutant human ASL were treated with varying concentrations of urea for 1 h at room temperature in assay buffer. ASL activity was then assessed as reported for **a**. Denaturation was reversible

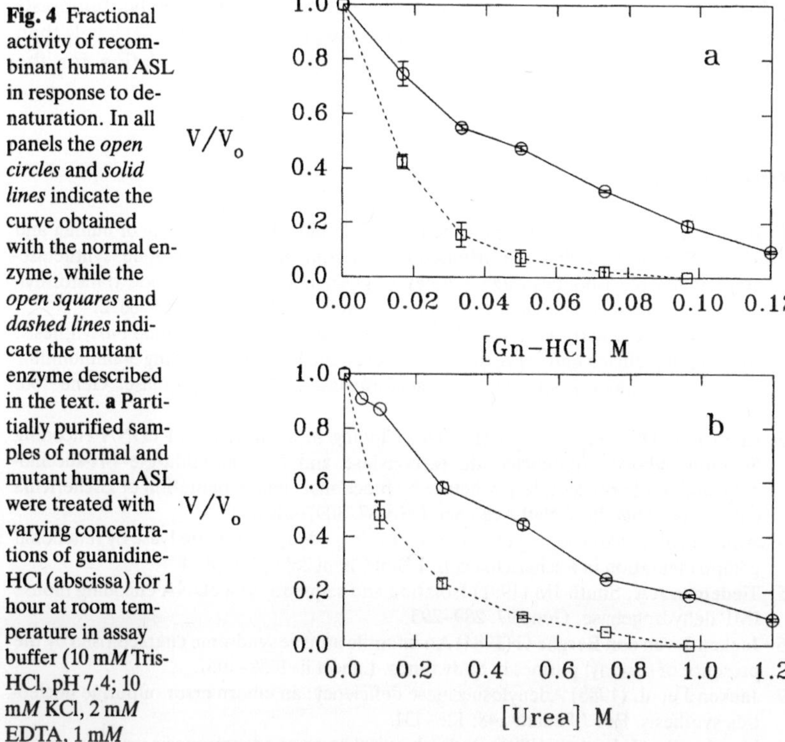



## Conclusions

We have shown that the ASL deficiency in a Moroccan family segregates with a point mutation which causes a Ser[413] → Pro substitution in the enzyme. Affected individuals in the family are homozygous for the point mutation, while all unaffected members of the family are heterozygous for the point mutation. The resultant recombinant protein is labile to two different denaturants. Free energy changes imply that the amino acid substitution disrupts a small number of hydrophobic interactions critical to active site structure. Cultured lymphoblasts from affected individuals in the family have lower steady state ASL levels than control lymphoblasts, suggesting an enhanced rate of degradation for the mutant enzyme.

# References


1. Aimi J, et al. (1990) De novo purine nucleotide biosynthesis: cloning of human and avian cDNAs encoding the trifunctional glycinamide ribonucleotide synthetase-aminoimidazole ribonucleotide synthetase-glycinamide ribonucleotide transformylase by functional complementation in E. coli Nucl Acids Res 18: 6665–6672.
2. Ni L, Guan K, Zalkin H, Dixon JE (1991) De novo purine nucleotide biosynthesis: cloning, sequencing and expression of a chicken PurH cDNA encoding 5-aminoimidazole-4-carboxamide-ribonucleotide transformylase-IMP cyclohydrolase. Gene 106: 197–205
3. Chen ZD, Dixon JE, Zalkin H (1990) Cloning of a chicken liver cDNA encoding 5-aminoimidazole ribonucleotide carboxylase and 5-aminoimidazole-4-N-succino-caroxamide ribonucleotide synthetase by functional complementation of Escherichia coli pur mutants. Proc Natl Acad Sci, USA 87: 3097–3101.
4. Aimi J et al. (1990) Cloning of a cDNA encoding adenylosuccinate lyase by functional complementation in Escherichia coli. J Biol Chem 265: 9011–9014
5. Tiedeman AA, Smith JM (1991) Isolation and sequence of a cDNA encoding mouse IMP dehydrogenase. Gene 97: 289–293.
6. Jaeken J, van den Berghe G (1984) An infantile autistic syndrome characterised by the presence of succinyl purines in body fluids. Lancet ii: 1058–1061.
7. Jaeken J et al. (1988) Adenylosuccinase deficiency: an inborn error of purine nucleotide synthesis. Eur J Pediatr 148: 126–131.
8. Van den Bergh F, et al. (1991) Radiochemical assay of adenylosuccinase: demonstration of parallel loss of activity toward both adenylosuccinate and succinylaminoimidazole carboxamide ribotide in liver of patients with the enzyme defect. Anal Biochem. 193: 287–291.
9. Barshop B, Alberts A, Gruber H (1989) Kinetic studies of mutant human adenylosuccinase. Biochim Biophys Acta 999: 19–23
10. Tabor S, Richardson CC (1985) A bacteriophage T7 RNA polymerase/promoter system for controlled exclusive expression of specific genes. Proc Natl Acad Sci, USA 82: 1074–1078

# V  Pyrimidine Metabolism

# 1 Dihydropyrimidinuria Presenting in Childhood with Severe Developmental Retardation

K. WARD, M. J. HENDERSON, H. A. SIMMONDS, J. A. DULEY, and P. M. DAVIES

## Introduction

Dihydropyrimidinase (5,6-dihydropyrimidine amidohydrolase; EC 3.5.2.2), is the second enzyme involved in the breakdown of the pyrimidine bases uracil and thymine and catalyses the degradation of dihydrouracil and dihydrothymine to β-ureidopropionic acid and β-ureidoisobutyric acid, respectively. The first case of dihydropyrimidinuria in humans was reported recently in an infant presenting with convulsions (Duran et al. 1991). A deficiency of dihydropyrimidinase was assumed from the accumulation and excretion in the urine of the substrates for the enzyme, dihydrouracil and dihydrothymine. A single defect of the pyrimidine catabolic pathway had been described previously, involving the precursor enzyme dihydropyrimidine dehydrogenase (DHPD: EC 1.3.1.2), in patients with a variety of neurological abnormalities Brockstedt et al. 1990).

This short report concerns the first case of dihydropyrimidinuria to be identified in the UK. The patient presented at six weeks with severe and sustained neurological deficits (Henderson et al., in press).

## Clinical History and Laboratory Investigations

This male child was the fourth child of healthy Pakistani parents believed to be first cousins. Although pregnancy and delivery were unremarkable and early development was normal, the baby presented with a febrile illness and seizures at 6 weeks and suffered frequent generalised seizures at approximately two monthly intervals thereafter. When first investigated at 2½ years of age, gross microcephaly and development retardation were evident, with signs of spastic quadriplegia. Jerky choreoform movements of the upper limbs were observed and the child could not sit without support. Apart from a respiratory tract infection, general examination revealed no other clinical abnormalities. Routine biochemical and haematological investigations, including amino acid analysis, were not abnormal.

A possible abnormality of pyrimidine metabolism was first suspected from the presence of the unusual components dihydrothymine and dihydrouracil in quantity in the urine during organic acid screening of trimethyl silyl derivatives of organic acids by gas chromatography-mass spectrometry (GC-MS), high-performance liquid chromatography (HPLC) with in-line diode-array analysis confirmed the presence of lesser amounts of thymine and uracil ($\sim$ 0.1 mmol/mmol creatinine). A deficiency of DHPD was excluded by the finding of normal activity in fibroblasts and lymphoblasts. No other abnormality of purine or pyrimidine metabolism was detected.

## Discussion

The excretion of the intermediates in the pyrimidine catabolic pathway, dihydrothymine and dihydrouracil, together with lesser amounts of thymine and uracil, by this infant is consistent with a deficiency of the enzyme dihydropyrimidinase. Identification of the specific metabolites excreted in this disorder by HPLC is difficult because the maximal UV absorbance of both compounds is below 230 nm. Their presence may be established by GC-MS or by HPLC by monitoring with diode-array and UV detection from 190 to 320 nm. Moderate amounts of uracil and thymine were also excreted by the patient reported by Duran et al. (1991).

The enzyme defect is, likewise, difficult to establish. Erythrocytes, leucocytes and fibroblasts are apparently devoid of activity, as are most other tissues. Liver appears obligatory for definitive proof of the presence or absence of activity. In this new patient with dihydropyrimidinuria, exclusion of a deficiency of DHPD in leucocytes and fibroblasts has been used as an indirect confirmation of diagnosis. Oral loading tests with uracil, dihydrouracil, thymine and dihydrothymine (1 mmol/kg) were used in the first case to distinguish the defect from DHPD deficiency (Duran et al. 1991).

The similarity of the findings in these unrelated infants, both from consanguineous kindreds is consistent with an autosomal recessive mode of inheritance. The molecular defect is unknown and the long-term prognosis is difficult to predict.

In contrast to the patient reported on by Duran et al. (1991), in whom no further seizures occurred after the initial presentation at 8 weeks and subsequent development was normal, this infant experienced frequent seizures for the first 2½ years of life and was clearly grossly retarded. He was treated with sodium valproate to good effect, but, unfortunately, has been lost to subsequent follow-up.

The differing severity of the clinical symptoms in the two reported patients suggest that, as with other genetic metabolic purine and pyrimidine disorders, phenotypic expression is broad. The recent description of the disorder coupled with the fact that the underlying biochemical defect is not readily identifiable by HPLC means it might not be so rare and could easily be missed.

# References

Brockstedt M, Jakobs C, Smit LME, van Gennip AH, Berger R (1990) A new case of dihydropyrimidine dehydrogenase deficiency. J Inher Metab Dis 13: 121–124

Duran M, Rovers P, de Bree PK, Schreuder CH, Beukenhorst H, Dorland L, Berger R (1991) Dihydropyrimidinuria: a new inborn error of pyrimidine metabolism. J Inher Metab Dis 14: 367–370

Henderson MJ, Ward K, Simmonds HA, Duley JA, Davies PM (1993) Dihydropyrimidinase deficiency presenting in infancy with severe developmental delay. J Inher Metab Dis (in press)

Van Gennip AH, Busch S, Scholten EG, Abeling NGMM (1992) Simple method for the quantitative analysis of dihydropyrimidines and N-carbamyl-β-amino acids in urine. Adv Exp Med Biol 309B: 15–19

# 2 The Clinical Aspects of Inherited Defects in Pyrimidine Degradation

M. Tuchman

## Introduction

Degradation pathways of pyrimidine bases are common to uracil, thymine, and the halogenated analogues of uracil and involve the same enzymes (Fig. 1). Cytosine nucleotides undergo deamination to form uracil nucleotides before entering the pyrimidine degradation pathway. Degradation of uracil and thymine occurs mainly in the liver although other tissues are also involved in this metabolic process (Levine et al. 1974; Naguib et al. 1985). The first degradative enzyme of this pathway is dihydropyrimidine dehydrogenase (DPD; EC 1.3.1.2.), an NADPH-dependent enzyme that reversibly reduces uracil and thymine forming dihydrouracil (DHU) and dihydrothymine (DHT), respectively. DPD enzymatic activity, which is considered rate limiting, is highest in the liver; however, it can be measured in most other tissues including white blood cells and cultured skin fibroblasts. The next degradative step involves enzymatic cleavage of the dihydropyrimidine ring by dihydropyrimidine amidohydrolase (DHPH; EC 3.5.2.2) followed by degradation of ureidopropionate and ureidoisobutyrate to from β-alanine and β-aminoisobutyrate, respectively, by the enzyme uriedopropionase (UP; EC 3.5.1.6). The latter two enzymes are expressed exclusively in liver. The β-alanine and β-aminoisobutyrate are then further metabolized to carbon dioxide and ammonia.

Pyrimidine base analogues including 5-fluorouracil (5-FU) and 5-fluorouridine (5-FUdR) are metabolized by the same pathway (Chaudhuri et al. 1958). These analogues are being used for chemotherapy in patients with various solid tumors. Therefore, blocks in the pyrimidine degradation pathway will prolong the half-life of these drugs enhancing their toxicity. The clinical and laboratory aspects of patients with defects in pyrimidine base degradation will be described in this report.

## Clinical Aspects

Patients with defects in pyrimidine base degradation have elevated concentrations of uracil and thymine in their body fluids (pyrimidinemia and pyrimidinuria). The first patient with such a defect was described by Berglund et al. (1979). They reported on a 2 year old child with a brain tumor (medulloblas-

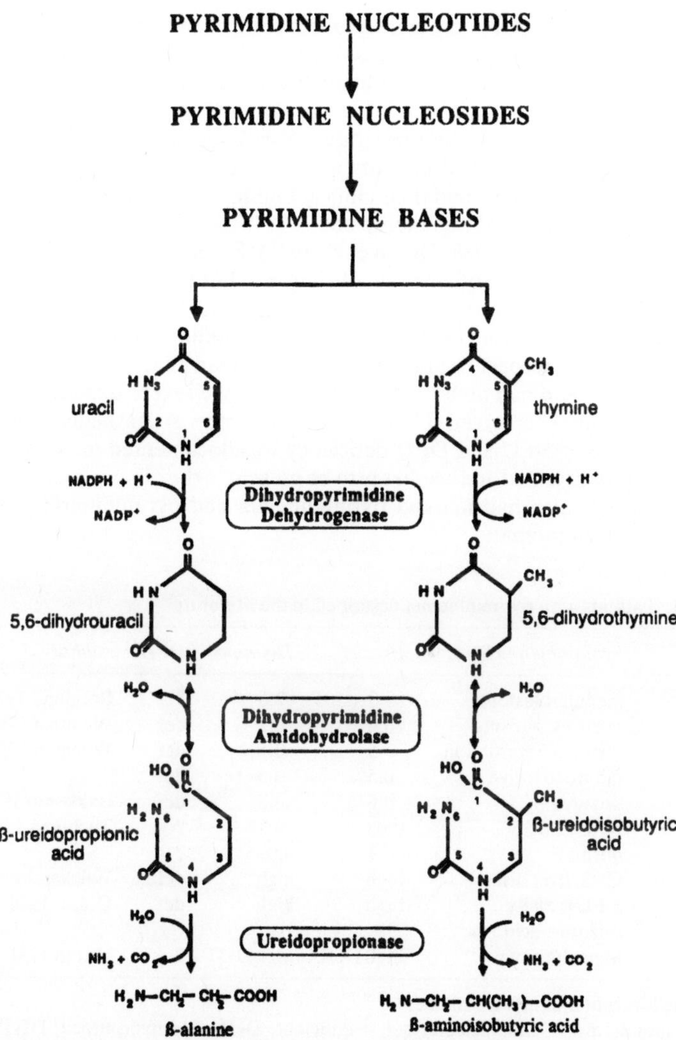



**Fig. 1.** Pyrimidine degradative pathways. Only the first enzyme of this pathway, the NADPH-dependent dihydropyrimidine dehydrogenase (DPD) is expressed in blood mononuclear cells and cultured skin fibroblasts. The subsequent two enzymes are expressed mainly in the liver. (From Tuchman et al. 1988)

toma) who excreted extremely high amounts of uracil and thymine in his urine. Other patients with malignant diseases usually have only mild elevations of urinary pyrimidines. Therefore, it is likely that the child had a DPD deficiency. His fibroblasts were assayed for DPD enzyme activity but the results were equivocal because a spectrophotometric assay with very low sensitivity was used (see below). Nine additional patients with pyrimidinuria and pyrimidinemia have since been described and their clinical findings are summarized in Table 1 (Wadman et al. 1984, 1985; Bakkeren et al. 1984; Tuchman et al. 1985; Wilcken et al. 1985, Diasio et al. 1988; Duran et al. 1991). Several of these patients were children with seizures or other brain dysfunction but this was not a consistent feature in all cases. Failure to thrive was found in one child and liver disease in another. Three adults from two families with pyrimidinuria were described who were normal with respect to brain function (Tuchman et al. 1985; Diasio et al. 1988). Two of them had breast cancer and developed severe toxicity including reversible coma when given 5-FU for chemotherapy. A markedly prolonged blood half-life of 5-FU and DPD deficiency was documented in one of these patients who was tested and was likely to be present in the other. A sibling of one of the patients also had marked pyrimidinemia and pyrimidinuria and was completely asymptomatic.

**Table 1.** Patients with pyrimidinemia described in the literature.

| Age | Symptoms and Signs | Uracil* | Thymine* | DPD | Author |
|---|---|---|---|---|---|
| 2y | medulloblastoma | 3 | 2.6 | ? | Berglund 1979 |
| 2y | seizures, autism | 0.4 | 0.2 | def | Wadman 1984 |
| 14y | seizures, retardation | 0.5 | 0.5 | def | Wadman 1985 |
| 15m | failure to thrive | 0.5 | 0.5 | def | |
| 3y | seizures | 0.6 | 0.2 | def | Bakkeren 1984 |
| 27y | 5-FU toxicity | 0.08 | 0.1 | ? | Tuchman 1985 |
| 32y | normal | 0.1 | 0.09 | ? | |
| newborn | CNS, liver disease | high | high | def | Wilcken 1985 |
| 40y | 5-FU toxicity | high | high | def | Diasio 1988 |
| 2mo | seizures, acidosis hypoglycemia | high DHU | high DHT | ? | Duran 1991 |

* Urine levels in mol/mol creatinine.
abbreviations: nd = undetectable, def = deficient, DHU = dihydrouracil, DHT = di-hydrothymine, CNS = central nervous system, 5-FU = 5-fluorouracil

Duran et al. (1991) described a 2 month old infant with seizures and acidosis who had elevated urinary uracil and thymine as well as their dihydro metabolites DHU and DHT. That child is likely to have DHPH deficiency based on his pyrimidine excretion patterns and results of loading studies (see below).

It is unknown whether pyrimidine degradation defects cause morbidity other than a definite predisposition for 5-FU toxicity which is well documented. The other clinical observations in these patients could either be related to the metabolic defect or be coincidental findings. The clinical spectrum is very wide, thus requiring more investigations into the cause and effect of these disorders.

## Laboratory Aspects

Accumulation of both uracil and thymine is the chemical marker of pyrimidine base degradation defects because both bases are metabolized by the same enzymes. Uracil and thymine are recovered by solvent extraction of organic acids and will therefore appear as abnormal peaks on the gas chromatogram of urinary organic acids. Excretion of elevated uracil without thymine can be seen in other disorders associated usually with orotic aciduria such as urea cycle defects (e.g., ornithine transcarbamylase deficiency). Except for the first patient, described by Berglund et al. (1979), who had pyrimidinuria in the range of 2.6–3.0 mol/mol creatinine, the urine excretion of uracil or thymine in the other patients ranged from 0.1 to 0.6 mol/mol creatinine (see Table 1). In normal adults, urinary uracil is around 0.002 mol/mol creatinine (less than 0.01 mol/mol in children) while thymine is usually undetectable in normal urine by routine methods (Tuchman et al. 1985). Wadman et al. (1984) detected urinary 5-hydroxymethyluracil, which is usually not detectable in the urine of normal individuals. This compound, the product of thymine hydroxylation, is not extractable with ethyl acetate and was detected by thin-layer chromatography (TLC). Blood levels of uracil and thymine are also markedly elevated in patients with DPD deficiency. Bakkeren et al. (1984) found levels of 24 and 19 µmol/l for uracil and thymine, respectively, in their patient. Tuchman et al. (1985) and Diasio et al. (1988) reported values of 6–13 µmol/l for uracil and 9–16 µmol/l for thymine, normal values being less than 0.5 and 0.1 µmol/l, respectively.

Patients with cancer rarely excrete more than twice the normal amount of pyrimidine bases in their urine (Berglund et al. 1979). Thus, patients with markedly elevated concentrations of both uracil and thymine as well as the presence of 5-hydroxymethyluracil in body fluids should be considered as having defective pyrimidine base catabolism.

Since several patients with pyrimidinemia and pyrimidinuria exhibited central nervous system symptoms, the concentrations of uracil and thymine in the cerebrospinal fluid (CSF) are of interest. Tuchman et al. (1985) and Diasio et al. (1988) found concentrations of pyrimidine bases in the CSF of their patients to be at least ten fold lower than in the plasma, whereas Bakkeren et al. 1984) found similar levels in blood and CSF of their patient. In all patients tested however, CSF levels of pyrimidine bases were abnormally elevated. The enhanced toxicity from 5-FU, especially to the central nervous system, described in patients with defects in pyrimidine base degradation may be explained by the prolonged blood half-life of the drug resulting possibly in higher drug levels over longer periods of time in the brain. Interference with normal brain RNA synthesis by 5-FU is likely to occur in such cases.

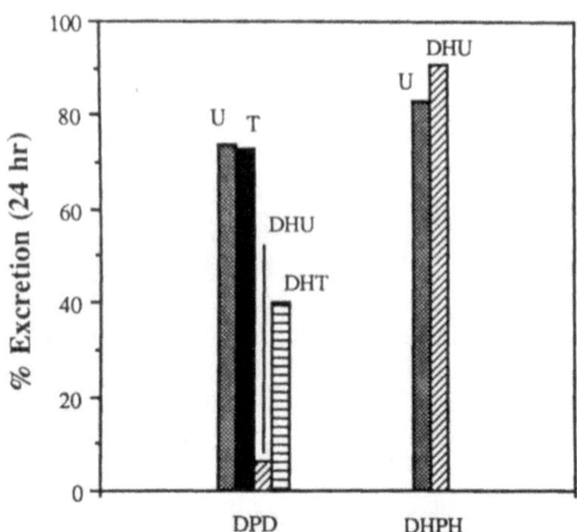



**Fig. 2.** Pyrimidine base loading in a patient with dihydropyrimidine dehydrogenase (DPD) deficiency and a patient with (dihydropyrimidine amidohydrolose) (DHPH) deficiency reported by Wadman et al. (1985). The figure illustrates the percentage of dose excreted within 24 h after a 1 mmol/kg load of each of the pyrimidine bases shown. *U*, uracil; T thymine; *DHU*, dihydrouracil, *DHT*, dihydrothymine

When DHU and DHT, the dihydro metabolites of uracil and thymine, are found in patients with pyrimidinuria and pyrimidinemia the diagnosis of DHPH is very likely. Recently, one patient with this finding has been described (Duran et al. 1991). Uracil and thymine concentrations were also elevated in that patient, but to a lesser degree than DHU and DHT. Loading studies described below should easily distinguish between DPD and DHPH deficiency.

Oral administration (1 mmol/kg) of uracil or thymine to normal individuals results in liver metabolism of these bases with very little (less than 10%) of the administered dose excreted unchanged in the urine (Wadman et al. 1984). In patients with DPD deficiency, however, most of the uracil ingested ($> 70\%$) was excreted unchanged in the urine reflecting deficient enzymatic conversion of uracil to DHU. In DPD deficient patients, less than 10% of DHU is found unchanged in the urine after an oral load of this compound, reflecting its normal further catabolism. On the other hand, in the patient with DHPH deficiency, oral load of either uracil or DHU results in excretion of mostly unchanged base in the urine, reflecting deficient cleavage of the dihydropyrimidine ring. Figure 2 illustrates the results of these loading tests in a patient with DPD deficiency and a patient with DHPH deficiency as described by Duran et al. (1991).

To confirm the diagnosis of DPD deficiency, enzymatic assays can be used in white blood cells or cultured skin fibroblasts (Bakkeren et al. 1984; de Abreu et al. 1986; Tuchman et al. 1989; Duran et al. 1991). The highest activity of DPD is found in the liver. Thus, a spectrophotometric assay which measures the oxidation of NADPH during the reduction of uracil or thymine, or an assay measuring the formation of uracil or thymine using the reverse reaction from dihydro substrates, are adequate for liver tissue. The spectrophotometric assays are not sensitive enough, however, to measure DPD activity in white blood cells or fibroblasts because the absorption of the blank is too high to reliably measure low enzymatic activity. The liver expresses enzymatic activities of all pyrimidine

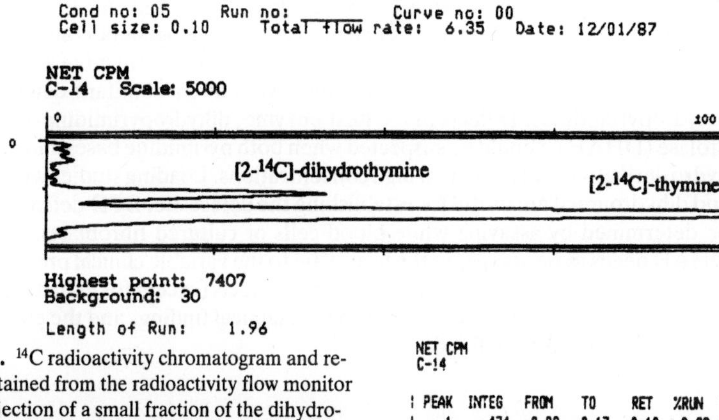



**Fig. 3A.** $^{14}C$ radioactivity chromatogram and report obtained from the radioactivity flow monitor after injection of a small fraction of the dihydropyrimidine dehydrogenase (DPD) reaction mixture into the high-performance liquid chromatography (HPLC). The radioactive peaks of $[2\text{-}^{14}C]$ thymine (substrate) and $[2\text{-}^{14}C]$ dihydrothymine (product) were separated by HPLC and their isotopic content quantitated by the radioactivity flow monitor. (From Tuchman et al. 1989)

NET CPM
C-14

| PEAK | INTEG | FROM | TO | RET | %RUN |
|---|---|---|---|---|---|
| 1 | 474 | 0.03 | 0.17 | 0.10 | 0.83 |
| 2 | 642 | 0.17 | 0.33 | 0.27 | 1.13 |
| 3 | 730 | 0.33 | 0.57 | 0.40 | 1.28 |
| 4 | 336 | 0.57 | 0.77 | 0.63 | 0.59 |
| 5 | 226 | 0.77 | 0.90 | 0.83 | 0.40 |
| 6 | 11325 | 0.90 | 1.30 | 1.10 | 19.93 |
| 7 | 42592 | 1.30 | 1.80 | 1.47 | 74.94 |
| 8 | 508 | 1.80 | 1.97 | 1.90 | 0.89 |
| - | 56833 | | | | TOTAL |

degradative enzymes and the whole pathway can therefore be screened for the conversion of radioactive uracil to radioactive carbon dioxide using liver homogenate. The most sensitive and specific in vitro assays for DPD activity are those employing a radioactive substrates and separation of the substrate from the product by TLC (Naguib et al. 1985), or preferably by HPLC combined with radioactivity flow monitoring as illustrated in Fig. 3 (Bakkeren et al. 1984, Tuchman et al. 1989). In a study of white blood cells from 45 adults, Tuchman et al. (1989) found DPD activity measured by radiochromatography to be $8.2 \pm 2.5$ (mean $\pm$ SD) nmol/mg protein/h with a range from 4.4 to 12.3 nmol/mg/h. The activity of DHPH is not detectable in white blood cells or fibroblasts and has to be measured in the liver.

## Summary

The pyrimidine bases uracil and thymine, as well as their fluorinated analogues, are degraded enzymatically through a sequence of steps beginning with reduction by DPD. Patients with inherited defects in pyrimidine base degradation have very elevated levels of both uracil and thymine (as well as 5-hydroxymethyluracil) in their urine, blood, and CSF. Ten patients with these defects have been described and their clinical presentations were variable ranging from normal health to cancer to severe central nervous disease and mental retardation. Members of two families with pyrimidinemia and pyrimidinuria due to

DPD deficiency have shown unusually severe toxicity when given 5-FU for cancer treatment confirming that DPD deficiency is a pharmacogenetic condition. The diagnosis of DPD deficiency is likely when both uracil and thymine are present in markedly increased levels in body fluids without concomitant detection of dihydropyrimidines. Defects in the next enzyme, dihydropyrimidine amidohydrolase (DHAH), should be suspected when both pyrimidine bases and their dihydro derivatives are present in high concentrations. Loading studies with uracil and dihydrouracil are useful for establishing the defect site. DPD deficiency can be determined by assaying white blood cells or cultured fibroblasts, whereas DHAH needs to be assayed in the liver. Due to the variable clinical presentation of patients with pyrimidine base degradation defects (except for enhanced 5-FU toxicity), the exact relationship between the clinical findings and the enzymatic defects remains to be defined.

# References


Bakkeren JAJM, de Abreu RA, Sengers RCA, Gabreëls FJM, Maas JM, Renier WO (1984) Elevated urine, blood and cerebrospinal fluid levels of uracil and thymine in a child with dihydrothymine dehydrogenase deficiency. Clin Chim Acta 140: 247–256

Berglund G, Greter J, Lindstedt S, Steen G, Waldenstrom J, Wass U (1979) Urinary excretion of thymine and uracil in a two-year-old child with a malignant tumor of the brain. Clin Chem 25: 1325–1328

Chaudhuri NK, Mukherjee KL, Heidelberger C (1958) Studies on fluorinated pyrimidines. VII – the degradative pathway. Biochem Pharm 1: 328–341

De Abreu RA, Bakkeren JAJM, Braakhekke J, Gabreëls FJM, Maas JM, Sengers RCA (1986) Dihydrothymine dehydrogenase deficiency in a family, leading to elevated levels of uracil and thymine. Adv Exp Med Biol 195A: 77–80

Diasio RB, Beavers, TL, Carpenter JT (1988) Familial deficiency of dihydropyrimidine dehydrogenase. Biochemical basis for familial pyrimidinemia and severe 5-fluorouracil-induced toxicity. J Clin Invest 81: 47–51

Duran M, Rover P, de Bree PK, Schreuder CH, Beukenhorst H, Dorland L, Berger R (1991) Dihydropyrimidinuria: a new inborn error of pyrimidine metabolism. J Inher Metab Dis 14: 367–370

Levine RL, Hoogenraad NJ, Kretchmer N (1974) A review: Biological and clinical aspects of pyrimidine metabolism, Pediatr Res 8: 724–734

Naguib FNB, el Kouni MH, Cha S (1985) Enyzmes of uracil catabolism in normal and neoplastic human tissues. Cancer Res 45: 5405–5412

Tuchman M, Stoeckeler JS, Kiang DT, O'Dea RF, Ramnaraine ML, Mirkin BL (1985) Familial pyrimidinemia and pyrimidinuria associated with severe fluorouracil toxicity. N Engl J Med 313: 245–249

Tuchman M, O'Dea RF, Ramnaraine MLR, Mirkin BL (1988) Pyrimidine base degradation in cultured murine C-1300 neroblastoma cells and in situ tumours. J Clin Invest 81: 425–430

Tuchman M, Roemeling RV, Hrushesky WAM, Sothern RB, O'Dea RF (1989) Dihydropyrimidine dehydrogenase activity in human blood mononuclear cells. Enzyme 42: 15–24

Wadman SK, Beemer FA, de Bree PK, Duran M, van Gennip AH, Ketting D, van Sprang FJ (1984) New defects of pyrimidine metabolism. Adv Exp med Biol 165A: 109–114

Wadman SK, Berger R, Duran M, de Bree PK, Stoker-de Vries SA, Beemer FA, Weits-Binnerts JJ, Penders TJ, van der Woude JK (1985) Dihydropyrimidine dehydrogenase deficiency leading to thymine-uraciluria. An inborn error of pyrimidine metabolism. J Inher Metab Dis 8 (suppl 2): 113–114

Wilcken B, Hammond J (1985) Dihydropyrimidine dehydrogenase deficiency – a further case. J Inher Metab Dis 8 (Suppl 2): 115–116

# 3  Dihydropyrimidine Dehydrogenase Deficiency: Biochemical and Genetic Basis

R. A. De Abreu

## Pyrimidine Disorders

The number of known inherited defects of pyrimidine metabolism is limited. Only four enzyme defects of pyrimidine metabolism have been described in association with inborn errors of metabolism:

- UMP-synthetase deficiency, generally ascribed as hereditary orotic aciduria
- Pyrimidine-5'-nucleotidase deficiency
- Dihydropyrimidine dehydrogenase deficiency
- Dihydropyrimidine amidohydrolase deficiency (a patient with this deficiency has been reported recently at the meeting of the Society of Inherited Metabolic Disease (SIMD) in Santa Fé [1])

The last three disorders involve defects of the pyrimidine degradative pathway (Fig. 1). Here, the dihydropyrimidine dehydrogenase deficiencies will be described in more detail.

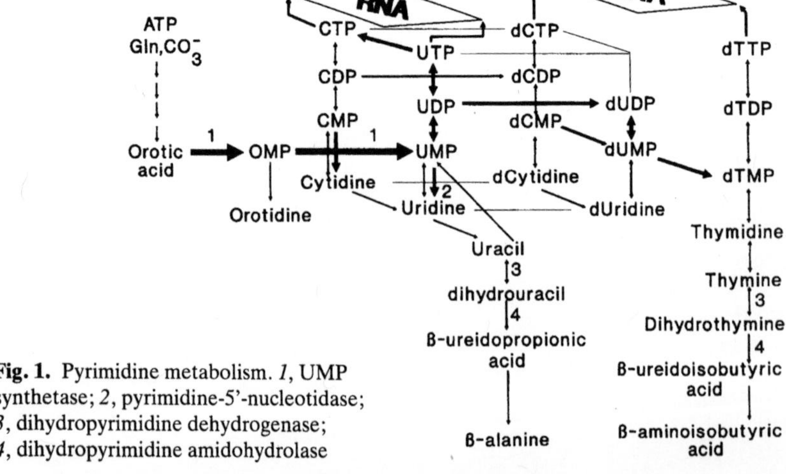


**Fig. 1.** Pyrimidine metabolism. *1*, UMP synthetase; *2*, pyrimidine-5'-nucleotidase; *3*, dihydropyrimidine dehydrogenase; *4*, dihydropyrimidine amidohydrolase

# Dihydropyrimidine Dehydrogenase

In mammalian tissue, dihydropyrimidine dehydrogenase catabolizes the first step in the degradation of pyrimidine bases. Dihydropyrimidine dehydrogenase is the rate-limiting enzyme in this catabolic pathway [2]. Both uracil and thymine are converted by the enzyme to dihydrouracil and dihydrothymine, respectively.

Dihydropyrimidine dehydrogenase is a NADPH-specific metalloflavoprotein. The enzyme from rat liver is a homodimer of approximately 220 kDa [3,4]. Recently, a NADPH-specific dihydropyrimidine dehydrogenase was purified from pig liver [5]. This enzyme is a homodimer of 206 Da. Each dimer contains two tightly associated molecules of FAD and FMN and 32 iron atoms per mole enzyme. The iron atoms are probably present in iron-sulfur centers. Data on the initial velocity and isotope exchange at equilibrium suggest a two-site ping-pong mechanism [6]. Product and dead-end inhibition patterns are consistent with a mechanism in which NADPH reduces the enzyme at site 1 and electrons are transferred to site 2 where uracil is reduced to dihydrouracil. The $Km$ values of both uracil (1 $\mu M$) and NADPH (7 $\mu M$) are low.

The effect of vitamin B2 deficiency on rat liver dihydropyrimidine dehydrogenase was investigated by Fujimoto [7]. The liver enzyme was diminished when rats were fed a vitamin B2 deficient diet for 5 weeks. Addition of exogenous flavin did not restore the enzyme activity. Therefore, endogenous flavin may regulate the enzyme its half-life or its synthesis.

# Dihydropyrimidine Dehydrogenase Deficiencies

Urinary excretion of excessive amounts of uracil and thymine are strong indications of a dihydropyrimidine dehydrogenase deficiency [8–16]. Elevated uracil and thymine levels can be confirmed in the plasma and cerebrospinal fluid of these patients [9, 12]. The enzyme defect was proven in fibroblasts [9], leukocytes [10], and liver [11].

In 1984 we reported on one of the first patients with a proven deficiency of dihydropyrimidine dehydrogenase [9]. More detailed clinical and laboratory data have been published elsewhere [12, 13].

In summary, the patient was the second child of consanguineous Dutch parents; she has an older sister and younger brother. During pregnancy her mother was treated for epilepsy with phenantoin and phenobarbitone. The delivery was at term. At 3 years of age, the child showed epileptic manifestations without fever. An electroencephalogram (EEG) demonstrated epileptic characteristics. Further examination revealed a right-handed microcephalic girl with normal height and weight.

Family consultation revealed several consanguineous relationships in the family. Both the mother and father of the patient were children of consanguineous parents. The grandmother and the mother presented with generalized tonic clonic seizures. EEGs of both showed generalized epileptic activity.

The father, the older sister, and younger brother of the propositus are normal and healthy individuals.

An accumulation of large amounts of uracil and thymine was detected in several body fluids of the patient, her mother, and younger brother. The elevated amounts of uracil and thymine from samples of the grandmother, father, and sister were less pronounced than in those of the patient (Table 1).

**Table 1.** Uracil and thymine concentrations in urine, plasma, and CSF from patients with dihydropyrimidine dehydrogenase deficiency

| Subject | Urine (mmol/mol creatine) | | plasma ($\mu$mol/liter) | | CSF ($\mu$mol/liter) | |
|---|---|---|---|---|---|---|
| | Uracil | Thymine | Uracil | Thymine | Uracil | Thymine |
| Propositus 1 | 594 | 187 | 24 | 19 | 25 | 15 |
| Mother | 263 | 89 | 15 | 13 | | |
| Father | 6 | 3 | 8.2 | 1.5 | | |
| Sister | | | 2.6 | 0.4 | | |
| Brother | | | 30 | 16 | | |
| Grandmother | | | 2.6 | 0.6 | | |
| Propositus 2 | 238 | 115 | 9 | 13 | 12 | 15 |
| Propositus 3 | 223 | 139 | 19 | 17 | 23 | 17 |
| Controls | <8 | nd | <0.2 | nd | <2.4 | nd |

nd, beneath the detection limit.

Increased levels of uracil and thymine seem to be a good reflection of the degree of enzyme deficiency. Patient 1, her brother, and mother have a complete deficiency. Her father and sister have a partial deficiency. No enzymatic data on the grandmother are available (Fig. 2).

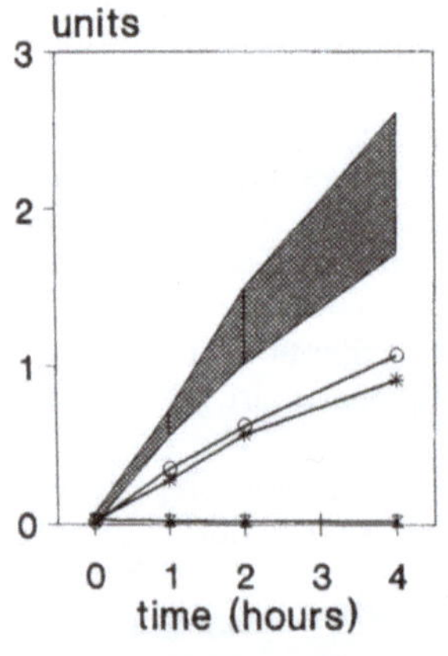


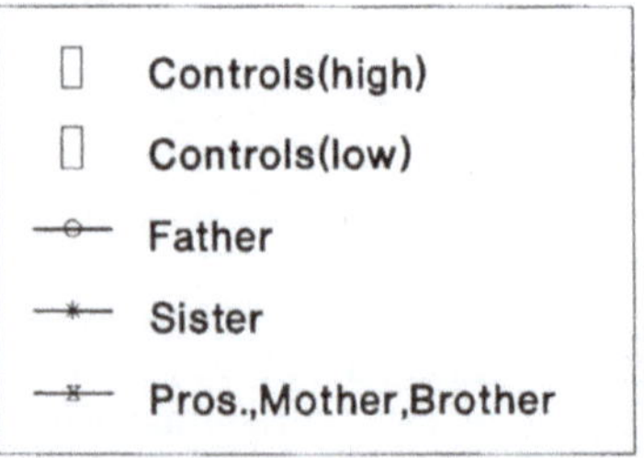


**Fig. 2.** Dihydropyrimidine dehydrogenase activity in cultured fibroblasts (nmoles dihydrothymine/mg protein)

Meanwhile, we have screened two other unrelated female patients with increased amounts of uracil and thymine in their body fluids (Table 1, propositus 2 and 3; Table 2, patients 4 and 5). A complete dihydropyrimidine dehydrogenase deficiency was also detected in fibroblasts of these patients.

Hitherto 13 patients with severe deficiency of dihydropyrimidine dehydrogenase have been documented ([19–16], AH van Gennip, personal communication) from which 11 are of Dutch origin. In all cases massive excretions of uracil (> 100 mmol/mol creatinine) and thymine (> 30 mmol/mol creatinine) were measured in the patients urine. The clinical presentation of these patients are summarized in Table. 2.

**Table 2.** Summary of diagnosed patients with dihydropyrimidine dehydrogenase deficiency

| Subjects Reference | Clinical symptoms |
| --- | --- |
| Patient 1, Dutch girl [9] | Epilepsy, microcephaly |
| Patient 2, mother of patient 1 [12, 13] | Epilepsy |
| Patient 3, brother of patient 1 [12, 13] | None |
| Patient 4, 18-year-old Dutch girl | Transient paresis |
| Patient 5, 14-year-old Dutch girl | Neurodegeneration |
| Patient 6, Dutch boy [10] | Epilepsy, autistic features, speech retardation |
| Patient 7, Dutch girl [10] | Mental retardation, dysplasia, epilepsy, solitary behavior, and petit mal |
| Patient 8, Turkish boy [10] | Mental retardation, microcephaly, severe growth retardation |
| Patient 9, Lebanese male baby [11] | Hepatomegaly, several episodes of generalized hypertonia and hyperflexia aspiration, ultimately died during one of these |
| Patient 10, Dutch girl [14] | Developmental problems |
| Patient 11, Dutch male baby [15] | Delay in psychomotoric development, microcephaly, generalized convulsions, axial imbalance |
| Patient 12, fetus [16] | Diagnosis during prenatal state |
| Patient 13, AH van Gennip, personal communication | Growth retardation |

Patient 12 ist the first case of prenatal diagnosis of dihydropyrimidine dehydrogenase deficiency [16]. Clearly elevated concentrations of thymine and uracil were measured in the amniotic fluid of the pregnancy at risk and termination of the pregnancy was decided upon. Patient 11 is a child of the pregnant woman. The deficiency was confirmed in the fetal liver.

# Dihydropyrimidine Dehydrogenase Deficiency in Connection with Malignancies

Apart from the patients mentioned in Table 2, four more patients have been reported on in connection with malignancies.

The first, a 2-year-old boy with medulloblastoma who excreted large amounts of uracil and thymine was described by Berglund [17]. However, severe deficiency of dihydropyrimidine dehydrogenase could not be demonstrated in this patient. Half the amount of normal dihydropyrimidine dehydrogenase activity was detected in his fibroblasts.

Recently, three cancer patients have been described with familial pyrimidinemia and severe neurotoxicity due to 5-fluorouracil. All three patients were females who underwent a left-modified radical mastectomy for infiltrating ductal carcinoma [18–20]. High levels of uracil and thymine were detected in urine and plasma from these patients. One patient died before a dihydropyrimidine dehydrogenase deficiency could be established [18]. Severe deficiency was established in peripheral blood mononuclear cells of the other two patients [20].

## Possible Pathophysiological Mechanisms

Although there is considerable variation in clinical presentation, neurological abnormalities are a common symptom among the reported patients. The pathophysiological mechanisms involved in the occurrence of neurological signs and symptoms in dihydropyrimidine dehydrogenase deficient patients are unclear and there is doubt whether a causal relationship exists between the deficiency and the symptoms. However, anticonvulsant effects of uridine in animals with experimentally induced seizures have been described indicating that pyrimidine compounds may play an important role in regulating nervous system activity [21]. It should also be considered that the three-step conversion of uracil via dihydrouracil and $N$-carbamoyl-β-alanine into β-alanine is the only pathway in mammalian tissues leading to the biosynthesis of β-alanine. Kautz and Schnackerz [22] have emphasized the importance of β-alanine as a putative neurotransmitter. Moreover, disturbances of pyrimidine metabolism by antimetabolites in cancer treatment are thought to be responsible for neurotoxicity presenting as seizures [23]. These data strongly suggest that inborn errors of pyrimidine metabolism might cause neurological symptoms.

The neurological abnormalities reported to develop in patients with familial pyrimidinemia and pyrimidinuria when treated with 5-fluorouracil can, at least partially, be explained by a greater bioavailability of 5-fluorouracil due to the decreased levels of catabolic enzyme. A circadian variation of dihydropyrimidine dehydrogenase activity in human peripheral blood mononuclear cells and a circadian variation of 5-fluorouracil plasma levels were observed in cancer patients receiving 5-fluorouracil by protracted continuous infusion. Under these conditions an inverse relationship between the circadian patterns of dihydropyrimidine dehydrogenase and 5-fluorouracil plasma level was noted, suggesting that an association exists between dihydropyrimidine dehydrogenase activity and 5-fluorouracil plasma concentrations [24].

## Mode of Inheritance

The results from studies on family members from our first patient with dihydropyrimidine dehydrogenase show evidence of partial deficiency in fibroblasts from the proband's father and sister and a complete deficiency in the fibroblasts from the mother and the brother (Fig. 2). This, coupled with the history of consanguinity in the family, suggests an autosomal recessive pattern of inheritance [9, 13]. This pattern of inheritance is consistent with results of family studies in other cases [10, 19, 20].

## References


1. Van Gennip AH, Rajnherc J, Busch S, van Cruchten A, Abeling NGGM (1991) Abnormal pyrimidine catabolism indicating dihydropyrimidinase deficiency in a patient with generalized convulsion. SIMD Meeting, Santa Fé
2. Weber G, Queener SF, Ferdinandus JA (1971) Control gene expression in carbohydrate, pyrimidine and DNA metabolism. In: Weber G (ed) Advance in enzyme regulation. Pergamon, New York, vol 9, pp 63–95
3. Shiotani T, Weber G (1981) Purification and properties of dihydrothymine dehydrogenase from rat liver. J Biol Chem 256: 219–224
4. Traut TW, Loechel S (1984) Pyrimidine catabolism: Individual characterization of the three sequential enzymes with a new assay. Biochemistry 23: 2533–2539
5. Podschun B, Wahler G, Schnackerz KD (1989) Purification and characterization of dihydropyrimidine dehydrogenase from pig liver. Eur J Biochem 185: 219–224
6. Podschun B, Cook PF, Schnackerz KD (1990) Kinetic mechanism of dihydropyrimidine dehydrogenase from pig liver. J Biol Chem 652: 12966–12972
7. Fujimoto S, Matsuda K, Kikugawa M, Kaneko M, Tamaki N (1991) Effect of vitamin B2 deficiency on rat liver dihydropyrimidine dehydrogenase activity. J Nutr Sci Vitamol 37: 89–98
8. Van Gennip AH, van Bree-Blom EJ, Wadman SK, Duran M, Beemer FA (1981) Liquid chromatography of urinary pyrimidines for the evaluation of primary and secondary abnormalities of pyrimidine metabolism. In: Hawk, GL, Champlin PB, Hutton RF, Mol C (eds) Biological/biomedical applications of liquid chromatography III. Dekker, New York, pp 285–296
9. Bakkeren JAJM, de Abreu RA, Sengers RC, Gabreëls FTM, Maas JM, Renier WO (1984) Elevated urine, blood and cerebrospinal fluid levels of uracil and thymine in a child with dihydrothymine dehydrogenase deficiency. Clin Chem Acta 140: 247–256
10. Berger R, Stoker-de Vries SA, Wadman SK, Duran M, Beemer FA, de Bree PK, Weits-Binnerts JJ, Penders TJ, van der Woude JK (1984) Dihydropyrimidine dehydrogenase deficiency leading to thymine-uraciluria. An inborn error of pyrimidine metabolism. Clin Chem Acta 141: 227–234

11. Wilcken B, Hammond J, Berger R, Wise G, James C (1985) Dihydropyrimidine dehydrogenase deficiency – A further case. J Inher Metab Dis 8, suppl 2: 115–116
12. De Abreu RA, Bakkeren JA, Braakhekke J, Gabreëls FJ, Maas JM, Sengers RC (1986) Dihydropyrimidine dehydrogenase deficiency in a family, leading to elevated levels of uracil and thymine. Adv Exp Med Biol 195A: 71–76
13. Braakhekke JP, Renier WO, Gabreëls FJ, De Abreu RA, Bakkeren JA, Sengers RC (1987) Dihydropyrimidine dehydrogenase deficiency, neurological aspects. J Neurol Sci 78: 71–77
14. Van Gennip AH, Bakker HD, Zoetekomer A, Abeling NGGM (1987) A new case of thymine-uraciluria. Klin Wochenschr 65, suppl. 8: 14
15. Brockstedt M, Jacobs C, Smit LME, van Gennip AH, Berger R (1990) A new case of dihydropyrimidine dehydrogenase deficiency. J Inher Metab Dis 13: 121–124
16. Jacobs C., Stellaard F, Smit LME, van Vugt JMG, Duran M, Berger R, Rovers P (1991) The first prenatal diagnosis of dihydropyrimidine dehydrogenase deficiency. Eur J Pediatr 150: 291
17. Berglund G, Greter J, Lindstedt S, Steen G, Waldenstroem J, Wass U (1979) Urinary excretion of thymine and uracil in a two-year-old child with a malignant tumor of the brain. Clin Chem 25: 1325–1328
18. Tuchman M, Stoeckeler JS, Kiang DT, O'Dea RF, Ramnaraine ML, Mirkin BL (1985) Familial pyridinemia and pyridinuria associated with severe fluorouracil toxicity. N Engl J Med 313: 245–249
19. Diasio RB, Beavers TL, Carpenter JT (1988) Familial deficiency of dihydropyrimidine dehydrogenase: biochemical bases for familial pyridinemia and severe 5-fluoro-uracil-toxicity. J Clin Invest 81: 47–51
20. Harries BE, Carpenter JT, Biasio RB (1991) Severe 5-fluoro-uracil toxicity secondary to dihydropyrimidine dehydrogenase deficiency. Cancer 68: 499–501
21. Roberts CA (1973) Anticonvulsant effects of uridine: comparative analysis of metrazol and penicillin induced foci, Brain Res 55: 291–308
22. Kautz J, Schnackerz KD (1989) Purification and properties of 5,6-dihydropyrimidine amido hydrolase from calf liver. Eur J Biochem 181: 431–435
23. Wiley RG, Gralla RJ, Casper ES, Kemeny N (1982) Neurotoxicity of the pyrimidine synthesis inhibitor N-phospho-acetyl-L-aspartate. Ann Neurol 12: 175–183
24. Harris BE, Song R, Soong S-J, Diasio RB (1990) Relationship between dihydropyrimidine dehydrogenase activity and plasma 5-fluorouracil levels with evidence for circadian variation of enzyme activity and plasma drug levels in cancer patients receiving 5-fluorouracil by protracted continuous infusion. Cancer Res 50: 197–201